John Gross Barnard

Analysis of Rotary Motion as Applied to the Gyroscope

John Gross Barnard

Analysis of Rotary Motion as Applied to the Gyroscope

ISBN/EAN: 9783744695206

Printed in Europe, USA, Canada, Australia, Japan

Cover: Foto ©ninafisch / pixelio.de

More available books at **www.hansebooks.com**

ANALYSIS

OF

ROTARY MOTION,

AS APPLIED TO

THE GYROSCOPE.

BY

MAJOR J. G. BARNARD, A. M.

NEW YORK:
D. VAN NOSTRAND, PUBLISHER,
23 MURRAY AND 27 WARREN STREETS.
1887.

PREFACE.

THE apparatus discussed here under the name Gyroscope was exhibited by Professor Walter R. Johnson, of the University of Pennsylvania, in 1831. It was then called the Rotascope, but it excited but little interest.

Professor Foucault, of France, brought it forward in 1855, and employed it as a means of making the earth's rotation visible to the eye. Since that time some form of Gyroscope (the name given it by Foucault) has formed a part of the philosophical apparatus for schools.

For some time the impression prevailed in the popular mind that the phenomena exhibited by the apparatus could not be explained by natural laws. This idea was perhaps strengthened by the name applied to it by Professor Olmstead, who called it "The Mechanical Paradox."

The following analytical exposition of

the motions of the Gyroscope was written by General (then Major) Barnard in 1858, for the *Journal of Education*. It was immediately reprinted in pamphlet form and was eagerly sought for by students of Analytical Mechanics. It yet remains the best treatise on this interesting apparatus.

As the former editions were long since exhausted, while the demand for the essay continued, it was considered advisable to republish it in its original form, first as a Magazine article and then as a volume of the Science Series.

G. W. P.

ANALYSIS OF ROTARY MOTION,
AS APPLIED TO
THE GYROSCOPE.

After reading most of the popular explanations of the above phenomenon given in our scientific and other publications, I have found none altogether satisfactory. While, with more or less success, they expose the more obvious features of the phenomenon and find in the force of gravity an efficient cause of horizontal motion, they usually end in destroying the foundation on which their theory is built, and leave an effect to exist *without a cause;* a horizontal motion of the revolving disk about the point of support is supposed to be accounted for, while the descending motion, which is the first and direct effect of gravity (and without which no horizontal motion can take place), is ignored or supposed to be entirely eliminated. Indeed, it is gravely stated as a distinguishing peculiarity of rotary motion, that, while gravity acting upon a non-rotating body causes it to descend vertically, the same force acting upon a rotary body causes it to *move horizontally.* *A tendency to descend* is supposed to produce the effect of an *actual descent;* as if, in mechanics, a mere tendency to motion ever produced any effect whatever without that motion actually taking place.

Whatever "mystification" there may be in analysis—however it may hide its results under symbols unintelligible save to the initiated, it is most certain that the greater portion of the physical phenomena of the universe are utterly beyond the grasp of the human mind without its aid. The mind can—indeed it *must*—search out the inducing causes, bring them together and adjust them to each other, each in its proper relation to the rest; but farther than that (at least in complicated phenomena) unaided, it cannot go. It cannot *follow* these causes in all their various actions and reactions and at a given instant of time bring forth the results.

This, analysis alone can do. *After* it has accomplished this, it indeed usually furnishes a clue by which to trace how the workings of known mechanical laws have conspired to produce these results. This clue I now propose to find in the analysis of rotary motion as applied to the gyroscope.

The analysis I shall present, so far as determining the equations of motions is concerned, is mainly derived from the works of Poisson (vide "Journal de l'Ecole Polytech." vol. XVI—Traité de Mécanique, vol. II, p. 162). Following his steps and arriving at his analytical results, I propose to develop fully their meaning, and to show that they are expressions not merely of a visible phenomenon, but that they contain within themselves the sole clue to its explanation; while they dispel all that is mysterious or paradoxical, and in reducing it to merely a " particular case " of the laws of "rotary motion," throw much light upon the significance and working of those laws.

Although not unfamiliar to mathematicians, it may not be uninteresting to those who have not time to go through the long preliminary study necessary to enable them to take up with Poisson this special investigation, or whose studies in mechanics have led them no farther than to the general equations of "rotary motion," found in text books, to show how the particular equations of the gyroscopic motion may be deduced.

In so doing I shall closely follow him; making, however, some few modifications for the sake of brevity and of avoiding the use of numerous auxiliary quantities not necessary to the limited scope of this investigation.

The general equations of rotary motion are (see Prof. Bartlett's "Analytical Mechanics," Equations (228), p. 170):

$$\left.\begin{aligned} C\frac{dv_z}{dt} + v_x v_y (B-A) &= L_1 \\ B\frac{dv_y}{dt} + v_x v_z (A-C) &= M_1 \\ A\frac{dv_x}{dt} + v_y v_z (C-B) &= N_1 \end{aligned}\right\} \quad (1)$$

In the above expressions the rotating body (of any shape) ABCD, Fig. 1, is supposed retained by the *fixed* point within or without its mass) O. Ox, Oy and Oz are the three co-ordinate axes, *fixed in space*, to which the motion of the body is referred. Ox_1, Oy_1, Oz_1, are the three *principal axes* belonging to the point O, and which, of course, partake of the body's motion. The position of the body at any instant of time is determined by those of the moving axes.

A, B and C express the several "moments of inertia" of the mass with reference, respectively, to the three principal axes Ox_1, Oy_1, Oz_1; N_1, M_1 and L_1 are the moments of the *accelerating forces*, and v_x, v_y, v_z, the *components of rotary*

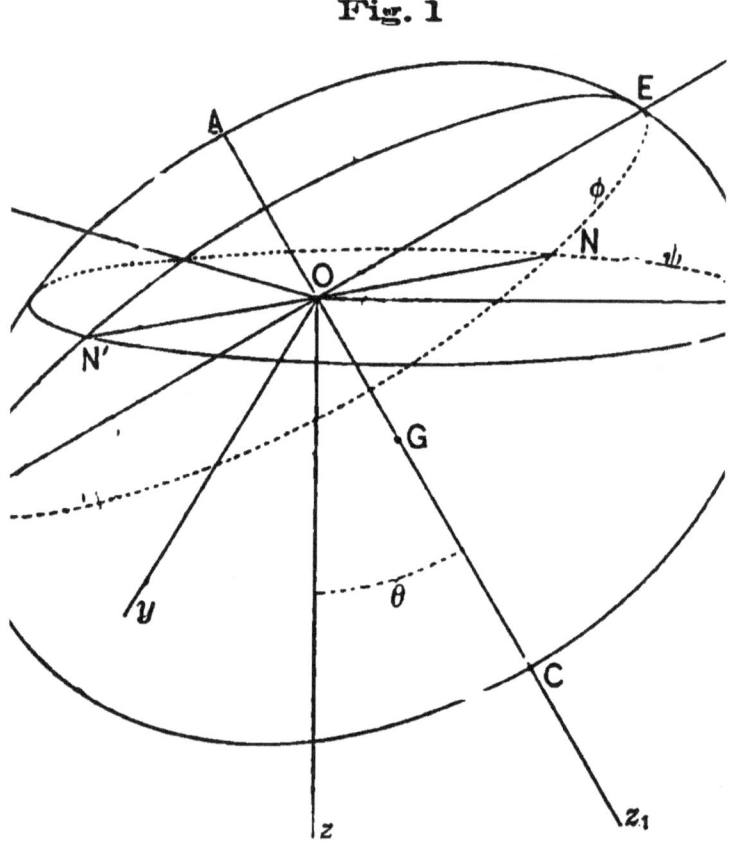

Fig. 1

velocity, all taken with reference to these same axes.

Like lineal velocities, velocities of rotation may be decomposed—that is, a rotation about any single axis may be considered as the resultant of components about other axes (which may always be reduced to three rectangular ones): and by this means, about whatever axis the body, at the instant we consider, may be revolving, its actual velocity and axis are determined by a knowledge of its components v_x, v_y, v_z, about the principal axes Ox_1, Oy_1, Oz_1, these components being, as with lineal velocities, equal to the resultant velocity multiplied by the cosine of the angles their several rectangular axes make with the resultant axis.

As the true axis and rotary velocity may continually vary, so the components v_x, v_y, v_z, in equations (1) are variable functions of the time.

For the purpose of determining the axes Ox_1, Oy_1 and Oz_1, with reference to the (fixed in space) axes Ox, Oy, Oz, three auxiliary angles are used.

If we suppose the moving plane of $x_1 y_1$, at the instant considered, to intersect the fixed plane of xy in the line NN' and call the angle $x\mathrm{ON} = \psi$, and the angle between the planes xy and $x_1 y_1$ (or the angle $z\mathrm{O}z_1) = \theta$, and the angle $\mathrm{NO}x_1 = \varphi$, (in the figure, these three angles are supposed *acute* at the instant taken) these three angles will determine the positions of the axes $\mathrm{O}x_1$, $\mathrm{O}y_1$, $\mathrm{O}z_1$, (and hence of the body) at any instant, and will themselves be functions of the time; and the rotary velocities v_x, v_y, v_z, may be expressed in terms of them and of their differential coefficients.

For this purpose, and for use hereafter in our analysis, it is necessary to know the values, in terms of φ, θ and ψ, of the cosines of the angles made by the axes $\mathrm{O}x_1$, $\mathrm{O}y_1$ and $\mathrm{O}z_1$ with the fixed axes $\mathrm{O}z$ and $\mathrm{O}y$.

These values are shown to be (vide Bartlett's Mech., p. 172)

$$\cos x_1 \mathrm{O}z = -\sin\theta \sin\varphi$$
$$\cos y_1 \mathrm{O}z = -\sin\theta \cos\varphi$$

$$\cos z_1 Oz = \cos \theta$$

$$\cos x_1 Oy = \cos \theta \cos \psi \sin \varphi \\ - \sin \psi \cos \varphi$$

$$\cos y_1 Oy = \cos \theta \cos \psi \cos \varphi \\ + \sin \psi \sin \varphi$$

$$\cos z_1 Oy = \sin \theta \cos \psi$$

The differential angular motions, in the time dt, about the axes Ox_1, Oy_1, Oz_1, will be $v_x dt$, $v_y dt$, and $v_z dt$. We may determine the values of these motions by applying the laws of composition of rotary motion to the rotations indicated by the increments of the angles θ, φ and ψ.

If θ and φ remain constant, the increment $d\psi$ would indicate that amount of angular motion about the axis Oz perpendicular to the plane in which this angle is measured. In the same manner $d\varphi$ would indicate angular motion about the axis Oz_1; while $d\theta$ indicates rotation about the line of nodes ON. In using these three angles, therefore, we actually refer the rotation to the three axes Oz, Oz_1, ON, of which one, Oz, is fixed in space, another, Oz_1, is fixed in and moves with the body, and the third, ON, is shifting in respect to both.

The angular motion produced around the axes Ox_1, Oy_1, Oz_1, by these simultaneous increments of the angles φ, θ and ψ, will be equal to the sum of the products of these increments by the cosines of the angles of these axes, respectively, with the lines Oz, Oz_1 and ON.

The axis of Oz_1 for example makes the angles θ, 0° and 90° with these lines, hence the angular motion $v_z dt$ is equal (taking the sum without regard to sign) to cos. $\theta d\psi + d\varphi$.

In the same manner (adding without regard to signs),

$$v_x dt = \cos. \ x_1 Oz d\psi + \cos. \ \varphi d\theta$$
and $v_y dt = \cos. \ y_1 Oz d\psi + \cos. \ (90° + \varphi) d\theta$.

But if we consider the motion about Oz_1 indicated by $d\varphi$, positive, it is plain from the directions in which φ and ψ are laid off on the figure, that the motion cos. $\theta d\psi$ will be in the reverse direction and negative, and since $\cos.\theta$ is positive $d\psi$ must be regarded as negative, hence

$$v_z dt = d\varphi - \cos.\theta d\psi$$

The first term of the value of $v_x dt$, $\cos. x_1 Oz d\psi$ [since $\cos. x_1 Oz$ ($= -\sin.\theta \sin.\varphi$) is negative and $d\psi$ is to be taken with the negative sign] is positive. But a study of the figure will show that the rotation referred to the axis Ox_1, indicated by the first term of this value, is the reverse of that measured by a positive increment of θ in the second, and hence, (as $\cos.\varphi$ is positive,) $d\theta$ must be considered negative. Making this change and substituting the values given of $\cos. x_1 Oz$, $\cos. y_1 Oz$, and for $\cos.(90° + \varphi), -\sin.\varphi$, we have the three equations

$$\left. \begin{array}{l} v_x dt = \sin.\theta \sin.\varphi d\psi - \cos.\varphi d\theta \\ v_y dt = \sin.\theta \cos.\varphi d\psi + \sin.\varphi d\theta \\ v_z dt = d\varphi - \cos.\theta d\psi \end{array} \right\} \quad (2)*$$

---

* To avoid the introduction of numerous quantities foreign to our particular investigation and a tedious analysis, I have departed from Poisson and substituted the above simple method of getting equations (2.), which is an instructive illustration of the principles of the composition of rotary motions.

The general equations (1) are susceptible of integration only in a few particular cases. Among these cases is that we consider, viz., that of a *solid of revolution* retained by a fixed point *in its axis of figure.*

Let the solid ABCD, Fig. 1, be supposed such a solid, of which Oz_1 is the axis of figure. It will be, of course, a principal axis, and any two rectangular axes in the plane, through O perpendicular to it, will likewise be principal. By way of determining them, let Ox_1 be supposed to pierce the surface in some arbitrarily assumed E point in this plane. Let G be the center of gravity (gravity being the sole accelerating force). The moments of inertia A and B become equal, and equations (1) reduce to

$$\left.\begin{array}{l} Cdv_z = 0 \\ Adv_y - (C-A)v_z v_x dt = \gamma a M g dt \\ Adv_x + (C-A)v_y v_z dt = -\gamma b M g dt \end{array}\right\} \quad (3)^*$$

---

* See Bartlett's Mech. Equations (225) and (118) for the values of $L_1 M_1 N_1$; in the case we consider the extraneous force P (of eq. 118) is g; the co-ordinates $x', y,$ of its point of application G (referred to the axes Ox_1 Oy_1, Oz_1,) are zero and $z^1 = OG = \gamma$: cosines of α, β and γ are a, b and c; hence $L_1 = 0$, $M_1 \gamma a M g$, $N_1 = -\gamma b M g$.

in which the distance OG of the point of support from the center of gravity is represented by γ, g is the force of gravity, M the mass and a and b stand for the cosines x_1Oz and y_1Oz and of which the values are

$$a = -\sin.\theta \sin.\varphi, \quad b = -\sin.\theta \cos.\varphi.$$

The first equation (3) gives by integration $v_z = n$, n being an arbitrary constant; it indicates that the rotation about the axis of figure remains always constant.

Multiplying the two last equations (3) by v_y and v_x respectively and adding the products, we get

$$A(v_y dv_y + v_x dv_x) = \gamma Mg(av_y - bv_x)dt.$$

From the values of a and b above, and from those v_x and v_y (equations 2) it is easy to find

$$(av_y - bv_x)dt = -\sin.\theta\, d\theta = d.\cos.\theta;$$

substituting this value and integrating and calling h the arbitrary constant

$$A(v_y^2 + v_x^2) = 2\gamma Mg \cos.\theta + h. \quad (a)$$

Multiplying the two last equations (3), respectively, by b and a and adding and reducing by the value just found of $d.\cos.\theta$ and of v_z, we get

$$A(bdv_y + adv_x) + (C-A)nd.\cos.\theta = 0 \quad (b)$$

Differentiating the values of a and b and referring to equations (2) it may readily be verified (putting for v_z its value n) that

$$db = (v_x \cos.\theta - an)dt$$
$$da = (bn - v_y \cos.\theta)dt$$

and multiplying the first by Av_y and the second by Av_x, and adding

$$A(v_y db + v_x da) = An(bv_x - av_y)dt$$
$$= -And.\cos.\theta.$$

Adding this to equation (b), we get

$Ad.(bv_y + av_x) + Cnd.\cos.\theta = 0$, the integral of which is

$A(bv_y + av_x) + Cn\cos.\theta = l$ (l being an arbitrary constant). $\qquad(c)$

Referring to equations (2) it will be found by performing the operations indicated that:

$$v_x{}^2 + v_y{}^2 = \sin.^2\theta \frac{d\psi^2}{dt^2} + \frac{d\theta^2}{dt^2}$$

$$bv_y + av_x = -\sin.^2\theta \frac{d\psi}{dt}$$

Substituting these values in equations (*a*) and (*c*), we get

$$C n. \cos.\theta - A \sin.^2\theta \frac{d\psi}{dt} = l$$

$$A\left(\sin.^2\theta \frac{d\psi^2}{dt^2} + \frac{d\theta^2}{dt^2}\right) = 2Mg\gamma \cos.\theta + h$$

If, at the origin of motion, the axis of figure is simply deviated from a vertical position by an arbitrary angle *a*, in the plane of *xz*, and an arbitrary velocity *n* is imparted about this axis alone; then v_x and v_y will at that instant be zero, $\theta = a$, and the substitution of these values in equations (*a*) and (*c*) will determine the values of the constants *l* and *h*.

$$h = -2Mg\gamma \cos. a$$
$$l = Cn \cos. a,$$

which, substituted in the above equations, make them

$$\left.\begin{aligned}\sin.^2\theta\frac{d\psi}{dt}&=\frac{Cn}{A}(\cos.\theta-\cos.a)\\ \sin.^2\theta\frac{d\psi^2}{dt^2}+\frac{d\theta^2}{dt^2}&=\frac{2Mg\gamma}{A}\\ &(\cos.\theta-\cos.a)\end{aligned}\right\} \quad (4)$$

These together with the last equation (2) which may be written, (substituting the value of v_z)

$$d\varphi = n dt + \cos.\theta d\psi \qquad (5)$$

will (if integrated) determine the three angles φ, θ and ψ in terms of the time t. They are therefore the differential equations of motion of the gyroscope.

Let NEE', (Fig. 1,) be a section of the solid by the plane $x_1 y_1$. This section may be called the *equator*. E being some fixed point in the equator (through which the principal axis Ox_1 passes), the angle φ is the angle EON.

If N is the *ascending node* of the equator—that is, the point at which E in its axial rotation *rises above* the horizontal plane, the angle φ must increase from N towards E—that is, $d\varphi$ (in equation 5) must be positive and (as the second term of its value is usually very small compared to the first) the angular velocity n must be positive. That being the case the value of $d\varphi$ will be exactly that due to the constant axial rotation ndt, augmented by the term cos. $\theta d\psi$, which is the projection on the plane of the equator of the angular motion $d\psi$ of the node. This term is an increment to ndt when it is positive, and the reverse when it is negative. In the first case, the motion of the node is considered *retrograde*—in the second, *direct*.

The first member of the second equation (4) being essentially positive, the difference cos. θ — cos. α must be always positive—that is, the axis of figure Oz, can never rise *above* its initial angle of elevation α. As a consequence $\dfrac{d\psi}{dt}$ [in first equation (4)] must be always positive. The node N, therefore, moves always in the direction in which ψ is laid off positively, and the motion will be direct or retrograde, with reference to the axial rotation, according as cos. θ is negative or positive—that is, as the axis of figure is above or below the horizontal plane. In either case the motion of the node in its own horizontal plane is always progressive in the same direction. If the rotation n were reversed, so would also be the motion of the node.

If this rotation n is zero, $\dfrac{d\psi}{dt}$ must also be zero and the second equation (4) reduces at once to the equation of the compound pendulum, as it should. Eliminating $\dfrac{d\psi}{dt}$ between the two equations (4) we get

$$\sin.^2\theta\frac{d\theta^2}{dt^2} = \frac{2\mathrm{M}g\gamma}{\mathrm{A}}\left[\sin.^2\theta - \frac{\mathrm{C}^2n^2}{2\mathrm{AM}\gamma g}(\cos.\theta - \cos. a)\right](\cos.\theta - \cos. a).$$

The length of the simple pendulum which would make its oscillations in the same time as the body (if the rotary velocity n were zero) is $\dfrac{A}{M\gamma}$.* If we call this λ and make for simplicity $\dfrac{C^2 n^2}{2A^2 g} = \dfrac{2\beta^2}{\lambda}$ the above equation becomes

$$\sin^2\theta \dfrac{d\theta^2}{dt^2} = \dfrac{2g}{\lambda}[\sin^2\theta - 2\beta^2(\cos.\theta - \cos.a)](\cos.\theta - \cos.a) \quad (6)$$

and the first equation (4) becomes

$$\sin^2\theta \dfrac{d\psi}{dt} = 2\beta\sqrt{\dfrac{g}{\lambda}}(\cos.\theta - \cos.a) \quad (7)$$

Equation (6) would, if integrated, give the value of θ in terms of the time; that is, the inclination which the axis of figure makes at any moment with the vertical; while eq. (7) (after substituting the ascertained value of θ) would give the value of ψ and hence determines the progressive movement of the body about the vertical Oz.

---

* The length of the simple pendulum is (see Bartlett's Mech., p. 252) $\lambda = \dfrac{k_1^2 + \gamma^2}{\gamma}$. The moment of inertia $A = M(k_1^2 + \gamma^2)$; hence $\dfrac{A}{M\gamma} = \lambda$.

These equations in the above general form, have not been integrated;* nevertheless they furnish the means of obtaining all that we desire with regard to gyroscopic motion, and in particular that self-sustaining power, which it is the particular object of our analysis to explain.

In the first place, from eq. (6), by putting $\dfrac{d\theta}{dt}$ equal to zero, we can obtain the maximum and minimum values of θ. This diff. coefficient is zero, when the factor cos. $\theta -$ cos. $a = 0$, that is, when $\theta = a$; and this is a *maximum*, for it has just been shown from equations (4) that θ cannot exceed a. It will be zero also and θ a *minimum*† when

$$\sin.^2\theta - 2\beta^2(\cos. \theta - \cos. a) = 0$$
or $\cos.\theta = -\beta^2 + \sqrt{1 + 2\beta^2\cos. a + \beta^4}$ (8)

---

* The integration may be effected by the use of elliptic functions; but the process is of no interest in this discussion.

† It is easy to show that this value of θ belongs to an actual minimum; but it is scarcely worth while to introduce the proof.

(The positive sign of the radical alone applies to the case, since the negative one would make θ a greater angle than a)

It is clear that (a being given) the value of θ depends on β alone, and that it can never become zero unless β is zero; and as long as the impressed rotary velocity n is not itself zero (however minute it may be), β will have a finite value:

Thus, however minute may be the velocity of rotation, it is sufficient to prevent the axis of rotation *from falling to a vertical position.*

The self-sustaining power of the gyroscope when very great velocities are given *is but an extreme case of this law.* For, if β is very great, the small quantity $1-\cos.^2 a$ may be subtracted from the quantity under the radical (eq. 8) without sensibly altering its value, which would cause that equation to become

$$\cos.\theta = \cos. a$$

That is, when the impressed velocity n, and in consequence β is very great, the minimum value of θ differs from its maximum a by an exceedingly minute quantity.

Here then is the result, analytically found, which so surprises the observer, and for which an explanation has been so much sought and so variously given. The revolving body, though solicited by gravity, *does not visibly fall.*

Knowing this fact, we may *assume* that the impressed velocity n is very great, and hence cos. $\theta -$cos. a exceedingly minute, and on this supposition, obtain integrals of equations (6) and (7), which will express with all requisite accuracy the true gyroscopic motion. For this purpose, make

$$\theta = a - u, \qquad d\theta = -du$$

in which the new variable u is always extremely minute, and is the angular descent of the axis of figure below its initial elevation.

By developing and neglecting the powers of u superior to the square, we have

$$\sin.^2 \theta = \sin.^2 a - u \sin. 2a + u^2 \cos. 2a*$$

$$\cos. \theta - \cos. a = u \sin. a - \tfrac{1}{2} u^2 \cos. a$$

substituting these values in eq. 6, we get

$$\sqrt{\frac{g}{\lambda}} dt = \frac{du}{\sqrt{2u \sin. a - u^2 (\cos. a + 4\beta^2)}}.\dagger$$

β having been assumed very great, cos. a

---

* By Stirling's theorem,

$$f(u) = U + U'\frac{u}{1} + U''\frac{u^2}{1.2}. \&c.,$$

in which U, U', U", &c, are the values of $f(u)$ and its different coefficients when u is is made zero.

Making $f(u) = \sin.^2(a-u)$, and recollecting that sin. $2u = 2 \sin. u \cos. u$ and cos. $2u = \cos.^2 u - \sin^2 u$, we get the value of sin.$^2\theta$; and making $f(u) = \cos.(a-u) - \cos. a$ the value in text of cos. θ − cos. a is obtained.

† Eq. 6 may be written

$$\frac{\lambda}{g} \frac{d\theta^2}{dt^2} = 2(\cos. \theta - \cos. a) - 4\beta^2 \frac{(\cos. \theta - \cos. a)^2}{\sin.^2 \theta.}$$

By substituting the values just found of $d\theta$, sin.$^2\theta$ and cos. θ − cos a and performing the operations indicated, neglecting the higher powers of u, (by which $\frac{(\cos. \theta - \cos. a)^2}{\sin.^2 \theta}$ reduces simply to u^2) and deducing the value $\sqrt{\frac{g}{\lambda}} dt$, the expression in the text is obtained.

may be neglected in comparison with $4\beta^2$ and the above may be written

$$\sqrt{\frac{g}{\lambda}}dt = \frac{du}{\sqrt{2u \sin. a - 4\beta^2 u^2}}. \qquad (d)$$

Integrating and observing that $u=o$, when $t=o$, we have

$$\sqrt{\frac{g}{\lambda}}.t = \frac{1}{2\beta}. \text{arc} \left\{ \cos. = 1 - \frac{4\beta^2 u}{\sin. a} \right\}$$

(*See Appendix, Note A.*)

$$u = \frac{\sin. a}{4\beta^2}\left(1 - \cos. 2\beta\sqrt{\frac{g}{\lambda}}.t\right)$$

or, (since $\cos. 2a = 1 - 2 \sin.^2 a$)

$$u = \frac{1}{2\beta^2} \sin. a \sin.^2 \beta\sqrt{\frac{g}{\lambda}}.t \qquad (9)$$

Putting $a-u$ in place of θ (equat. 7) neglecting square of u, we get

$$\frac{d\psi}{dt} = \frac{1}{\beta}\sqrt{\frac{g}{\lambda}}. \sin.^2 \beta\sqrt{\frac{g}{\lambda}}.t \qquad (10)$$

(*See Appendix, Note B.*)

from which, observing that $\psi = 0$, when $t = 0$

$$\psi = \frac{1}{2\beta}\sqrt{\frac{g}{\lambda}}.t - \frac{1}{4\beta^2}\sin.\left(2\beta\sqrt{\frac{g}{\lambda}}.t\right) \quad (11)$$

These three expressions (9), (10), (11), represent the vertical angular depression—the horizontal angular velocity—and the extent of horizontal angular motion of the axis of figure after any time t.*

The first two will reach their respective maxima and minima when $\sin.\beta\sqrt{\frac{g}{\lambda}}t = 1$ and $= 0$; or when $t = \frac{\pi}{2\beta}\sqrt{\frac{\lambda}{g}}$ and $t = \frac{\pi}{\beta}\sqrt{\frac{\lambda}{g}}$.

These values of t in equation (11) give

$$\psi = \frac{\pi}{4\beta^2} \qquad \psi = \frac{\pi}{2\beta^2}$$

---

* The assumption that $\psi = 0$ when t is zero supposes that the initial position of the node coincides with the fixed axis of x. In my subsequent illustrations and analysis I suppose the initial position to be at 90° therefrom, which would require to the above value of ϕ, the constant ½π to be added. The horizontal angular motion of the axis of figure is the same as that of the node.

Hence, counting from the commencement of motion, when t, u, $\dfrac{d\psi}{dt}$ and ψ are all zero, we have the following series of corresponding values of these variables

$$t=\frac{\pi}{2\beta}\sqrt{\frac{\lambda}{g}},\ u=\frac{1}{2\beta^2}\sin.\ a\frac{d\psi}{dt}=\frac{1}{\beta}\sqrt{\frac{g}{\lambda}}$$

$$\psi=\frac{\pi}{4\beta^2}$$

which correspond to the moment of greatest depression, when u and $\dfrac{d\psi}{dt}$ are maxima and

$$t=\frac{\pi}{\beta}\sqrt{\frac{\lambda}{g}},\ u=0,\ \frac{d\psi}{dt}=0,\ \psi=\frac{\pi}{2\beta^2},$$

when, it appears (u being the zero), the axis of figure has regained its original elevation and the horizontal velocity is destroyed.

All these values are (owing to the assumed large value of β) very minute. If we suppose the rotating velocity $n=100\ \pi$ or 100 revolutions per second, the maximum of u (with an instrument of ordinary proportions) would be a fraction of a minute of arc, and the period of undulation but a fraction of a second.

Hence the horizontal motion about the point of support will be exceedingly slow compared with the axial rotation of the disk expressed by n.

If, in equations (9) and (10), we increase t indefinitely we will have but a repetition of the series of values already found, they being recurring functions of the time.

. We see, then, the revolving body *does not*, in fact, maintain a uniform unchanging elevation, and move about its point of support at a uniform rate, (as it appears to do). But the axis of figure generates what may be called a *corrugated cone*, and any point of it would describe an undulating curve (Fig. 2) whose superior culminations a, a', a'', &c., are *cusps* lying in the same horizontal plane, and whose sagittae cb, $c'b'$, &c., are to the amplitudes aa', a', a'' &c., as $\dfrac{\sin. a}{2\beta^2} : \dfrac{\pi}{2\beta^2} :: \sin. a : \pi$. If the initial elevation a is 90°,

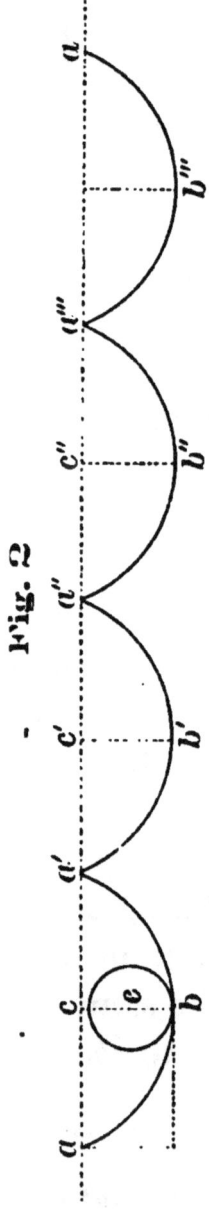

Fig. 2

this ratio is *as the diameter to the circumference of the circle:* a property which indicates the *cycloid.*

Assuming $a = 90°$ and sin. $a = 1$, equations (9) and (10) will give, by elimination of $\sin.^2\beta\sqrt{\frac{g}{\lambda}t}$,

$$\frac{d\psi}{dt} = 2\beta\sqrt{\frac{g}{\lambda}u} \qquad dt = \frac{d\psi}{2\beta\sqrt{\frac{g}{\lambda}u}},$$

substituting this value in eq. (*d*), we get

$$d\psi = \frac{2\beta u\, du}{\sqrt{2u - 4\beta^2 u^2}} = \frac{u\, du}{\sqrt{\frac{1}{2\beta^2}u - u^2}}$$

the differential equation of the cycloid generated by the circle whose diameter is $\frac{1}{2\beta^2}$

In this position of the axis, both the angles u and ψ are arcs of great circles described by a point of the axis of figure at a units distance from O, and owing to their minuteness may be considered as rectilinear co-ordinates.

If a is not 90°, the sagittae $bc = \dfrac{1}{2\beta^2}$ sin. a; but then, while the angular motion ψ is the same, the arc described by the same point of the axis will be that of a *small circle*, whose actual length will likewise be reduced in the ratio of 1 : sin.a. The curve is therefore a cycloid in all circumstances; and the axis of figure moves as if it were attached to the circumference of a minute circle whose diameter is $\dfrac{1}{2\beta^2}$sin. a, which rolled along the horizontal circle, a a' a'', about the vertical through the point of support.

The center e of this little circle moves with uniform velocity. The *first term* of the value of ψ (equation 11) is due to this uniform motion; it may be called the *mean precession*.

The second term is due to the circular motion of the axis about this center, and combined with the corresponding values of u, constitutes what may be called the *nutation*.

These cycloidal undulations are so minute—succeed each other with such rapidity (with the high degreees of velocity usually given to the gyroscope), that they are entirely lost to the eye, and the axis seems to maintain an unvarying elevation and move around the vertical with a uniform slow motion.

It is in omitting to take into account these minute undulations that nearly all popular explanations fail. They fail, in the first place, because they substitute, in the place of the real phenomenon, one which is purely imaginary and *inexplicable*, since it is in direct variance with fact and the laws of nature;—and they fail, because these undulations—(great or small, according as the impressed rotation is small or great) furnish the only true clue to an understanding of the subject.

The fact is, that the phenomenon exhibited by the gyroscope which is so striking, and for which explanations are so much sought, is only a *particular and extreme phase* of the motion expressed by equations (6) and (7)—that the self-sustaining power is not absolute, but one of degree—that, however minute the axial rotation may be, the body never will fall quite to the vertical;—however great, it cannot sustain itself without any depression.

I have exhibited the undulations, as they exist with high velocities—when they become minute and nearly true cycloids; with low velocities they would occupy (horizontally) a larger portion of the arc of a semi-circle, and reach downward approximating, more or less nearly, to contact with the vertical; and, *finally*, when the rotary velocity is zero, their cusps are in diametrically opposite points of the horizontal circle, while the curves resolve themselves into vertical circular arcs which coincide with each other, and the vibration of the pendulum is exhibited. All these varieties of motion, of which that of the pendulum is one extreme phase and the gyroscopic another, are embraced in equations (6) and (7) and exhibited by varying β from 0 to high values, though (wanting general integrals to these equations) we cannot determine, except in these extreme cases, the exact elements of the undulations. The minimum value of θ may, however, always be determined by equation (8).

If we scrutinize the *meaning* of equations (6) and (7), it will be found that they represent, the first, the horizontal angular component of the velocity of a point at units distance from O, and the second the actual velocity of such point.*

---

*In more general terms equations (4) express. the first, that the *moment of the quantity of motion* about the fixed vertical axis Oz remains always constant; the *second* that the living forces generated in the body (over and above the *impressed* axial rotation) are exactly what is *due to gravity through the height, h.*

Both are expressions of truths that might have been anticipated; for gravity cannot increase the moment of the quantity of motion about an *axis parallel to itself;* while its power of generating living force by working through a given height, cannot be impaired.

Had we considered ourselves at liberty to assume them, however, the equations might have been got without the tedious analysis by which we have reached them.

For $\sin.\theta \frac{d\psi}{dt}$ is the horizontal, and $\frac{d\theta}{dt}$ the vertical, component of this velocity. Calling the first v_h, and the second v_v, and the resultant v_s, and calling $\cos.\theta - \cos.a$ (which is the true height of fall), h, those equations may be written

$$v_h = \frac{Cn}{A} \frac{h}{\sin.\theta} \qquad (e)$$

$$(v_h^2 + v_v^2) = v_s^2 = \frac{2g}{\lambda} h \qquad (f)$$

This velocity v_s (as a function of the height of fall) is exactly that of *the compound pendulum,* and is *entirely independent of the axial rotation n.* Hence (as we might reasonably suppose) rotary motion has no power to impair the work of gravity *through a given height*, in generating velocity; but it does have power to *change the direction of that velocity.* Its effect is precisely that of a material undulatory curve, which, deflecting the body's path from vertical descent, finally directs it upward, and causes its velocity to be destroyed by the same forces which generated it.

And it may be remarked, that, were the cycloid we have described *such a material curve*, on which the axis of the gyroscope rested, without friction and *without rotation*, it would travel along this curve by the effect of gravity alone (the velocity of descent on the downward branch carrying it up the ascending one), with *exactly the same velocity* that the rotating disk does, through the combined effects of gravity *and* rotation.

Equation (*a*) expresses the horizontal velocity produced by the rotation.

If we substitute its value in the second we may deduce

$$v_v \text{ or } \frac{d\theta}{dt} = \sqrt{\frac{2g}{\lambda}h - \frac{C^2 n^2}{A^2} \frac{h^2}{\sin^2 \theta}}$$

If we take this value at the commencement of descent, *and before any horizontal velocity is acquired* (making h indefinitely small), the second term under the radical may be neglected, and the first increment of descending velocity becomes $\sqrt{\frac{2g}{\lambda}h}$, precisely what is due to gravity, and *what it would be were there no rotation.*

Hence the popular idea that a rotating body offers any *direct* resistance to a change of its plane, is unfounded. It requires as little exertion of force (in the direction of motion) to move it from one plane to another, as if no rotation existed; and (as a corollary) as little expenditure of work.

But deflecting forces are developed, by angular motion given to the axis, and normal to its direction, which are very sensible, and are mistaken for *direct* resistances. If the extremity of the axis of rotation were confined in a vertical circular groove, in which it could move without friction; or if any similar fixed resistance, as a material vertical plane, were opposed to the *deflecting* force, the rotating disk would vibrate in the vertical plane as if no rotation existed. Its equation of motion would become that of the compound pendulum, $\frac{d\theta}{dt}=\sqrt{\frac{2g}{\lambda}h}$. What then is the resistance to a change of plane of rotation so often alluded to and described. A *misnomer* entirely.

The above may be otherwise established. If in equations (3) we introduce in the second member an indeterminate horizontal force, g', applied to the center of gravity, parallel to the fixed axis of y, and contrary to the direction in which, in our figure, we suppose the angle ψ to increase, the projections of this force on the axes Ox_1, Oy_1, will be $a' g'$ and $b' g'$ and the last two of these equations will become (calling cosines $x_1 Oy$ and $y_1 Oy_1$, a' and b',)

$$A dv_y - (C-A)nv_x\, dt = \gamma M(ag + a'g')dt$$
$$A dv_x + (C-A)nv_y dt = -\gamma M(bg + b'g')dt$$

Multiplying the first by v_y and the second by v_x and adding

$$A(v_y dv_y + v^x dv_x) = \gamma M[g(av_y - bv_x)dt + g'(a'v_y - b'v_x)dt]$$

But $(av_y - bv_x)dt$ has been shown to be $= d.\cos.\theta$,—and by a similar process it may be shown that $(a'v_y - b'v_x)dt = d.(\sin.\theta\cos.\psi)$. (For values of a' and b' as before.)

Let us suppose now that the force g' is such that the axis of the disk may be always maintained in the plane of its initial position xz. The angle ψ would always be 90°, $d\psi = 0$, and $d.(\sin.\theta\cos.\psi) = 0$. That is, the co-efficient of the new force g' becomes zero; and the integral of the above equation is as before,

$$A(v_y^2 + v_x^2) = 2\gamma Mg\cos.\theta + h.$$

But the value of $v_y^2 + v_x^2$ likewise reduces (since $\dfrac{d\psi}{dt} = 0$) to $\dfrac{d\theta^2}{dt^2}$ and the above becomes the equation of the compound pendulum (g) $\dfrac{d\theta^2}{dt^2} = \dfrac{2\gamma Mg}{A}\cos.\theta + h = \dfrac{2g}{\lambda}$ $(\cos.\theta - \cos.a)$, (h being determined.) This is the principle just before announced, that, with a force so applied as to prevent any *deflection* from the plane in which gravity tends to cause the axis to vibrate, the motion would be precisely *as if no axial rotation existed.*

To determine the force of g'; multiply the first of preceding equations by b, and the second by a, and add the two, and add likewise $A(v_y db + v_x da = -A\, nd$ cos. θ, and we shall get

$$Ad(bv_y + av_x) + C\, nd\, \cos.\, \theta = \gamma Mg' (a'b - ab')dt.$$

By referring to the values of a, a', b, b', and performing the operations indicated and making cos. $\psi = o$, sin. $\psi = 1$, the above becomes,

$$Ad(bv_y + av_x) + C\, nd\, \cos.\, \theta = \gamma Mg' \sin.\, \theta\, dt.$$

But the value of $(bv_y + av_x)$ becomes zero when $\dfrac{d\psi}{dt} = o$. Hence,

$$g' = \frac{C\, nd\, \cos.\, \theta}{\gamma M\, \sin.\, \theta\, dt} = -\frac{Cn\, d\theta}{\gamma M\, dt} *$$

---

* The effect of gravity is to diminish θ and the increment $d\theta$ is negative in the case we are considering. Hence the negative sign to the value of g', indicating

The second factor $\frac{d\theta}{dt}$ is the *angular velocity* with which the axis of rotation is moving.

Hence calling v_s that angular velocity, *the value of the deflecting force, g'* may be written (irrespective of signs),

$$g' = \frac{C}{\gamma M} n v_s. \qquad (h)$$

that is, it is directly proportional *to the axial rotation n,* and *to the angular velocity* of the axis of that rotation. By putting for C, Mk^2 (in which k is the distance from the axis at which the mass M, if concentrated, would have the moment of inertia, C,) the above takes the simple form

$$g' = \frac{k^2}{\gamma} n v_s.$$

In the case we have been considering above, in which g' is supposed to coun-

---

that the force is in the direction of the *positive* axis of *y*, as it should, since the tendency of the node is to move in the reverse direction.

teract the deflecting force of axial rotation, the angular velocity v_s or $-\dfrac{d\beta}{dt}$ (equation g) is equal to $\sqrt{\dfrac{2g}{\lambda}(\cos.\theta - \cos.a)}$

But in the case of the *free* motion of the gyroscope, this deflecting force combines with gravity to produce the observed movements of the axis of figure.

If, *therefore*, we *disregard* the axial rotation and consider the body simply as fixed at the point O, and acted upon, at the center of gravity, by two forces — one of gravity constant in intensity and direction — the other, the *deflecting* force due to an axial rotation n, whose variable intensity is represented by $\dfrac{C}{\gamma M} nv_s$, and whose direction is always normal to the plane of motion of the axis; we ought, introducing these forces, and making the axial rotation n zero, in general equations (3), to be able to deduce therefrom the identical equations (4) which express the motion of the gyroscope.

This I have done; but as it is only a verification of what has previously been said, I omit in the text the introduction of the somewhat difficult analysis.

(*See Appendix, Note C.*)

Equation (5) becomes (in the case we consider), by integration;

$$\varphi = nt + \psi \cos. a$$

which, with the values of u and ψ already obtained, determines completely the position of the body at any instant of time.

Knowing now not only the exact nature of the motion of the gyroscope, but the direction and intensity of the forces which produce it, it is not difficult to understand why such a motion takes place.

Fig. 1 represents the body as supported by a point *within* its mass; but the analysis applies to any position, in the axis of figure, within or without; and Figs. 3 and 4 represent the more familiar cir-

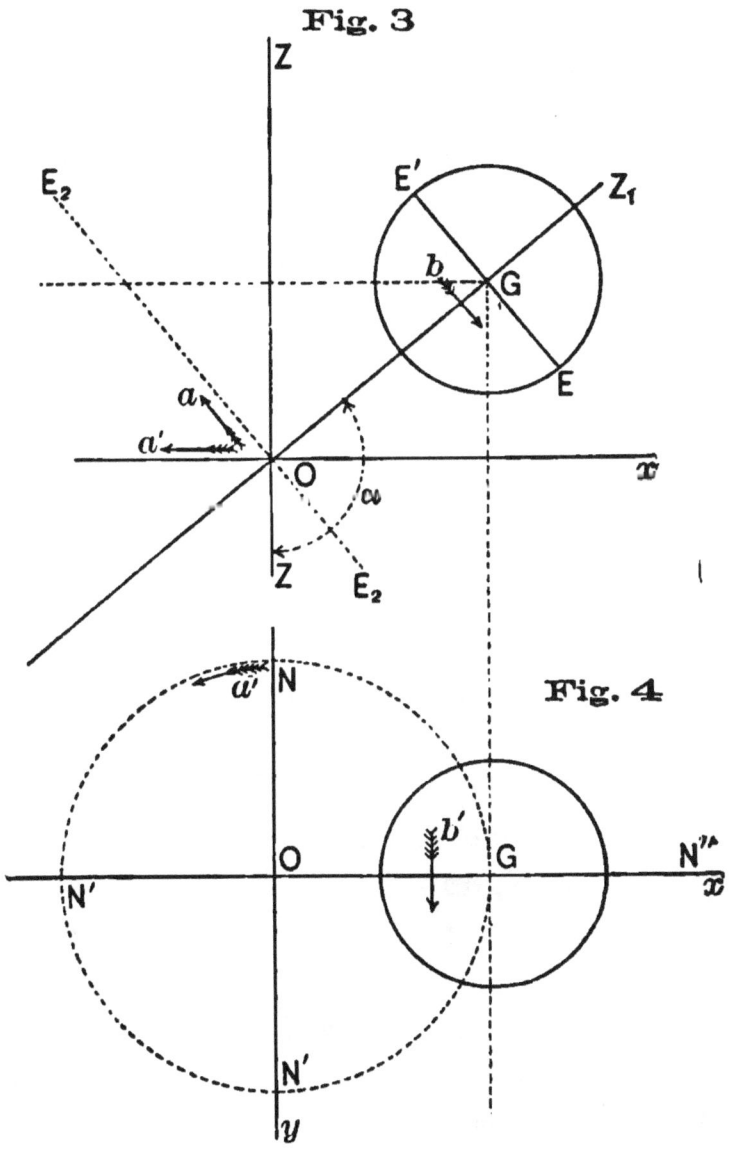

Fig. 3

Fig. 4

cumstances under which the phenomenon is exhibited.

Let the revolving body be supposed (Fig. 3, vertical projection), for simplicity of projection, an exact *sphere*, supported by a point in the axis prolonged at O, which has an initial elevation a greater than 90°. Fig. 4. represents the projection on the horizontal plane xy the initial position of the axis of figure (being in the plane of xz) is projected in Ox.

Ox, Oy, Oz, are the three (fixed in space) co-ordinate axes, to which the body's position is referred.

In this position, an initial and high velocity n is supposed to be given about the axis of figure Oz_1, so that the visible portions move in the direction of the arrows b, b', and the body is left subject to whatever motion about its point of support O, gravity may impress upon it. Had it no axial rotation, it would immediately fall and vibrate according to the known laws of the pendulum. Instead of which, while the axis maintains (apparently) its elevation *a*, it moves slowly around the vertical O*z*, receding from the observer, or from the position ON″ towards ON.

It is self-evident that the first *tendency* (and as I have likewise proved, the first effect) of gravity is to cause the axis Oz_1 to descend vertically, and to generate vertical *angular velocity*. But with this angular velocity, the *deflecting* force proportional to that velocity and normal to its direction, is generated, which pushes aside the descending axis from its vertical path. But as the direction of motion changes, so does the direction of this force—always preserving its perpendicularity. It finally acquires an intensity and upward direction adequate to neutralize the downward action of gravity; but the *acquired downward velocity* still exists and the axis *still* descends, at the same time acquiring a constantly increasing horizontal component, and with it a still increasing upward deflecting force. At length the descending component of velocity is entirely destroyed—the path of the axis is horizontal; the deflecting force due to it acts directly contrary to gravity, which it exceeds in intensity, and hence causes the axis to commence rising. This is the state of things at the point *b* (Fig. 2). The axis has descended the curve *ab*, and has acquired a velocity due to its *actual* fall *ad*; but this velocity has been deflected to a *horizontal direction*. The *ascent* of the branch *ba'* is precisely the converse of its descent. The *acquired* horizontal velocity impels the axis horizontally, while the deflecting force due to it (now at its maximum) causes it to commence ascending. As the curve bends upward, the normal direction of this force opposes itself more and more to the horizontal, while gravity is equally counteracting the vertical velocity. As the *horizontal* velocity at *b* was due to a fall through the height *ad*, so, through the medium of this deflecting force, it is just capable of restoring the *work* gravity had expended and *lifting* the axis back to its original elevation at *a'*, and the cycloidal undulation is completed, to be again and again repeated, and the axis of figure, performing undulations too rapid and too minute to be perceived, moves slowly around its point of support.

Referring to Fig. 3, the *equator* of the revolving body (a plane perpendicular to the axis of figure and *through the fixed point* O,) would be an imaginary plane $E_1 E_2$. Its intersection with the horizontal plane of xy would be the line of nodes N,N'. In the position delineated, the progression of the nodes is *direct*. For, at the *acending* node N, any point in the imaginary plane of the equator (suppoosed to revolve with the body) would move *upwards* in the direction of the arrow *a*, while the node moves in the *same direction from* O (of the arrow *a'*). Were the axis of figure below the horizontal plane, (Fig. 5) the upward rotation of the point would be from O to E_2 (as the arrow *a*), while the progression of the node (in the same direction as before as the arrow *a'*) would be the reverse, and the motion of the node would be *retro-*

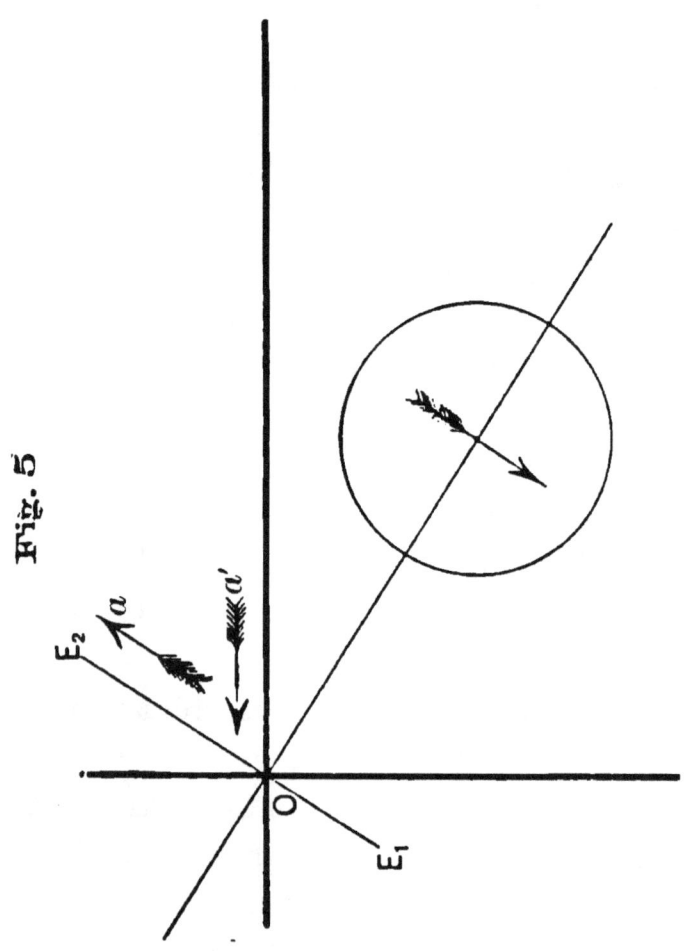

grade—yet in both cases the same in space.

As the deflecting force of rotary motion is the sole agent in diverting the vertical velocity produced by gravity from its downward direction, and in producing these paradoxical effects; and as the foregoing analysis, while it has determined its value, has thrown no light upon its origin, it may be well to inquire how this force is created.

Popular explanations have usually turned upon the deflection of the *vertical* components of rotary velocity by the vertical angular motion of the axis produced by gravity. In point of fact, however *both* vertical and horizontal components are deflected, one as much as the other; and the simplest way of studying the effects produced, is to trace a vertical projection of the path of a point of the body under these combined motions. For this purpose conceive the mass of the revolving disk concentrated in a single ring of matter of a radius k due to its moment of inertia $C = Mk^2$ (see Bartlett Mech. p.

178), and, for simplicity, suppose the angular motion of the axis to take place around the center figure and of gravity G.

Let AB be such a ring (supposed perpendicular to the plane of projection) revolving about its axis of figure GC, while the axis turns *in the vertical plane* about the same point G. Let the rotation be such that the visible portion of the disk moves upward through the semi-circumference, from B to A, while the axis moves downward through the angle θ to the position GC'. The point B, by its *axial rotation* alone, would be carried to A; but the plane of the disk, by simultaneous movement of the axis, is carried to the position A'B', and the point B arrives at B' instead of A, through the curve projected in BGB'. The equation of the projection, in circular functions, is easily made; but its general character is readily perceived, and it is sufficient to say that it passes through the point G,—that its tangents at B and B' are perpendicular to AB and A'B',—and that its concavity throughout its whole length turned to *the right.* The point A descends on the other, or remote side of the disk, and makes an exactly similar curve AGA′ with its concavity reversed.

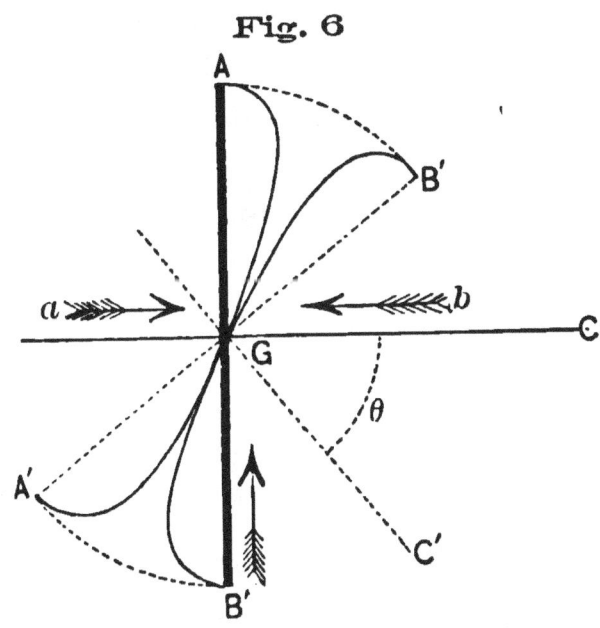

Fig. 6

The *centrifugal* forces due to the deflections of the vertical motions are normal to the concavities of these curves; hence, on the side of the axis *towards* the eye, they are to the *left,* and on the opposite or further side, to the *right,* (as the arrows *b* and *a.*) Hence the joint effect is to press the axis GC from its vertical plane CGC', horizontally and towards the eye. Reverse the direction of axial rotation and the curves AA' and BB' will be the same except that AA' would be on the *near* and BB' on the *remote* side of the axis GC, and the direction of the resulting pressure will be reversed.

A projection on the horizontal plane would likewise illustrate this deflecting force and show at the same time that there is *no resistance in the plane of motion of the axis*, and that the whole effect of these deflections of the paths of the different material points, is a mere *interchange of living forces between the different material points of the disk;* but it is believed that the foregoing illustration is sufficient to explain the *origin* of this force, whose measure and direction I have analytically demonstrated.

It may be remarked, however, that the intensity of the force will evidently be directly as the velocities *gained and lost* in the motion of the particles from one side of the axis to the other; or as the *angular velocity of the axis*, and as the distance, k, of the particles from that axis. It will also be as the *number of particles* which undergo this gain and loss of living force in a given time; or *as the velocity of axial rotation*. Considered as applied normally at G to produce rotation about *any* fixed point O in the axis, its intensity will evidently be *directly* as the arm of lever k, and *inversely* as the distance of G from O (γ). Hence the measure of this force already found, from analysis,

$$g' = \frac{k^2}{\gamma} n v_s.$$

In the foregoing analysis, the entire ponderable mass is supposed to partake of the impressed rotation about the axis of figure Oz_1; and such must be the case im order that the results we have arrived at may rigidly apply. Such, however, cannot be the case in practice. A portion of the instrument must consist of mountings which do not share in the rotation of the disk. It is believed the analysis will apply to this case by simply including the *whole mass*, in computing the moment of inertia A and the mass M, while the moment C represents, as before, that of *the disk alone*.

In this manner it would be easy to calculate what *amount of extraneous* weight (with an *assumed* maximum depression u) the instrument would sustain, with a given velocity of rotation.

The analogy between the minute motions of the gyroscope and that grand phenomenon exhibited in the heavens,—the "precession of the equinoxes"—is often remarked. In an ultimate analysis, the phenomena, doubtless, are identical; yet the immediate causes of the latter are so much more complex, that it is difficult to institute any profitable comparison.

At first sight the undulatory motion attending the precession, known as "nutation" (nodding) would seem identical with the undulations of the gyroscope. But the identity is not easily indicated; for the earth's motion of nutation is mainly governed by the moon, with whose cycles it coincides; and the solar and lunar precessions and nutations are so combined, and affected by causes which do not enter into our problem, that it is vain to attempt any minute identification of the phenomena, without reference to the difficult analysis of celestial mechanics.

On a preceding page I said that a horizontal motion of the rotating disk around its point of support, without descending undulations, was at variance with the laws of nature. This assertion applied, however, only to the actual problem in hand, in which no other external force than gravity was considered, and no other initial velocity than that of axial rotation.

Analysis shows, however, that an initial *impulse* may be applied to the rotating disk in such a way that the horizontal motion shall be absolutely without undulation. An initial horizontal angular velocity, such as would make its corresponding deflective force equal to the component of gravity, g sin. θ, would cause a horizontal motion *without* undulation.

If the axial rotation n, as well as the horizontal rotation, is communicated by an impulsive force, analysis shows that it may be applied in *any plane* intersecting the horizontal plane *in the line of nodes;* but if applied in the plane of the equator (where it can communicate nothing but an *axial* rotation n), or in the horizontal plane, its intensity must be infinite.

My announced object does not carry me further into the consideration of the gyroscope than the solution of this peculiar phenomenon, which depends solely upon, and is so illustrative of, the laws of rotary motion.

If I have been at all successful in making this so often explained subject more intelligible—in giving clearer views of some of the supposed effects of rotation, it has been because I have trusted solely to the *only* safe guide in the complicated phenomena of nature, *analysis*.

Appendix.—Note A.

$$\frac{du}{\sqrt{2u \sin. a - 4\beta^2 u^2}}$$ may be put in the form

$$\frac{2\beta}{\sin. a} \cdot \frac{\frac{\sin. a}{4\beta^2} du}{\sqrt{2u \frac{\sin. a}{4\beta^2} - u^2}}$$

Call $\frac{\sin. a}{4\beta^2} =$ R, and the integral of the 2d factor of the above is the arc whose radius is R and versed sine is u; or whse cosine is R$-u$, or it is R *times* the arc whose cosine $1 - \frac{u}{R}$ with radius unity. Substituting the value of R in the integral and multiplying by the factor $\frac{2\beta}{\sin. a}$ we get the value of $\sqrt{\frac{g}{\lambda}} t$, of the text.

Note B.

In eq. (7) if we divide both members by $\sin.^2 \theta$, and, in reducing the fraction $\dfrac{\cos. \theta - \cos. a}{\sin.^2 \theta}$, use the values already found and neglect the *square*, as well as higher powers u, (which may be done without sensible error, owing to the minuteness of u, though it could not be done in the foregoing values of dt and t, since the co-efficient $4\beta^2$ in those values, is reciprocally great, as u is small) the quotient will be simply $\dfrac{u}{\sin. a}$

Substituting the value of u and dividing out $\sin. a$, we get the value of $\dfrac{d\psi}{dt}$ in the text.

The integral of $\sin.^2 \beta \sqrt{\dfrac{g}{\lambda}} t\, dt$ results from the formula $\int \sin.^2 \varphi\, d\varphi = \frac{1}{2}\varphi - \frac{1}{4} \sin. 2\varphi$, easily obtained by substituting for $\sin.^2 \varphi$, its value $\frac{1}{2} - \frac{1}{2} \cos. 2\varphi$.

Note C.

To introduce these forces in Eq. (3) I observe, first, that as both are applied at G (in the axis Oz_1) the moment L_1 is still zero and the *first* eq. becomes, as before

$$Cdv_z = 0 \text{ or } v_z = \text{const.}$$

And as we disregard the impressed axial rotation, we make this constant (or v_z) zero.

The deflecting force $\dfrac{Cn}{\gamma M}v_s$ (taken with contrary sign to the *counteracting* force just obtained) resolves itself into two components $\dfrac{Cn}{\gamma M}\dfrac{d\theta}{dt}$ and $-\dfrac{Cn}{\gamma M}\dfrac{d\psi}{dt}$ sin. θ, the first in a horizontal, the second in a vertical plane, and both normal to the axis of figure.

The second is opposed to gravity, whose component normal to the axis of figure is g sin. θ.

Hence we have the two component forces (in the directions above indicated),

$$\text{M.}\frac{Cn}{\gamma \text{M}}\frac{d\theta}{dt} \text{ and } \text{M}\left(g-\frac{Cn}{\gamma \text{M}}\frac{d\psi}{dt}\right)\sin, \theta.$$

These moments with reference to the axes of y_1 and x_1 will be

$$-\sin.\varphi\gamma\text{M}\left(g-\frac{Cn}{\gamma\text{M}}\frac{d\psi}{dt}\right)\sin.\theta-$$
$$\cos.\varphi\gamma\text{M}\frac{Cn}{\gamma\text{M}}\frac{d\theta}{dt}, \text{ and}$$

$$\cos.\varphi\gamma\text{M}\left(g-\frac{Cn}{\gamma\text{M}}\frac{d\psi}{dt}\right)\sin.\theta-$$
$$\sin.\varphi\gamma\text{M}\frac{Cn}{\gamma\text{M}}\frac{d\theta}{dt}$$

Hence equations (3) (making v_z zero, and putting for M_1 and N_1 the above values, and recollecting the values of a and b, become

$$A dv_y = a_\gamma M g dt - $$
$$a C n \frac{d\psi}{dt} dt - C n \cos. \varphi \frac{d\theta}{dt} dt$$
$$A dv_x = -b_\gamma M g dt + $$
$$b C n \frac{d\psi}{dt} dt - C n \sin. \varphi \frac{d\theta}{dt} dt$$

$\Bigg\}\ i$

Multiplying the equations severally by v_y and v_x, adding and reducing, we get

$$A(v_y\, dv_y + v_x\, dv_x) = \gamma M g d. \cos. \theta - $$
$$C n \frac{d\psi}{dt} d. \cos. \theta - C n d\, \theta\, (v_y \cos. \psi + v_x \sin. \varphi)$$

But $v_y \cos. \varphi + v_x \sin. \varphi$ will be found equal to $\sin. \theta \frac{d\psi}{dt}$ (by substituting the values of v_y and v_x); hence the two last terms destroy each other, and the above equation becomes identical with equation (*a*) from which the 2d eq. (4) is deduced.

Multiplying the 1st equation (*i*) by cos. φ and the second by sin. φ and adding, we get

$$A(\cos. \varphi dv_y + \sin. \varphi dv_x) = -C n d\, \theta.$$

By differentiating the values of v_y and v_x, performing the multiplications, and substituting for $d\varphi$ its value, cos. $\theta d\psi$, (proceeding from the 3d equation (2) when $v_z = 0$), the above becomes

$$A\left(\sin.\theta \frac{d^2\psi}{dt^2} + 2\cos.\theta \frac{d\psi}{dt}\frac{d\theta}{dt}\right) = -Cn\frac{d\theta}{dt}.$$

Multiplying both members by sin. θdt, and integrating, the above becomes

$$\sin.^2\theta \frac{d\psi}{dt} = \frac{Cn}{A}\cos.\theta + l;$$

the same as the 1st equation (4) when the value of the constant l is determined.

*** *Any book in this Catalogue sent free by mail on receipt of price.*

VALUABLE
SCIENTIFIC BOOKS,

PUBLISHED BY

D. VAN NOSTRAND,

23 MURRAY STREET AND 27 WARREN STREET, N. Y.

ADAMS (J. W.) Sewers and Drains for Populous Districts. Embracing Rules and Formulas for the dimensions and construction of works of Sanitary Engineers. Second edition. 8vo, cloth $2 50

ALEXANDER (J. H.) Universal Dictionary of Weights and Measures, Ancient and Modern, reduced to the standards of the United States of America. New edition, enlarged. 8vo, cloth 3 50

ATWOOD (GEO.) Practical Blow-Pipe Assaying. 12mo, cloth, illustrated 2 00

AUCHINCLOSS (W. S.) Link and Valve Motions Simplified. Illustrated with 37 wood-cuts and 21 lithographic plates, together with a Travel Scale and numerous useful tables. 8vo, cloth 3 00

AXON (W. E. A.) The Mechanic's Friend: a Collection of Receipts and Practical Suggestions Relating to Aquaria—Bronzing—Cements—Drawing—Dyes—Electricity—Gilding—Glass-working—Glues—Horology—Lacquers—Locomotives—Magnetism—Metal-working—Modelling—Photography—Pyrotechny—Railways—Solders—Steam-Engine—Telegraphy—Taxidermy—Varnishes—Waterproofing, and Miscellaneous Tools, Instruments, Machines, and Processes connected with the Chemical and Mechanic Arts. With numerous diagrams and wood-cuts. Fancy cloth 1 50

BACON (F. W.) A Treatise on the Richards Steam-Engine Indicator, with directions for its use. By Charles T. Porter. Revised, with notes and large additions as developed by American practice; with an appendix containing useful formulæ and rules for engineers. Illustrated. Third edition. 12mo, cloth 1 00

BARBA (J.) The Use of Steel for Constructive Purposes; Method of Working, Applying, and Testing Plates and Brass. With a Preface by A. L. Holley, C.E. 12mo, cloth. $1 50

BARNES (Lt. Com. J. S., U. S. N.) Submarine Warfare, offensive and defensive, including a discussion of the offensive Torpedo System, its effects upon Iron-Clad Ship Systems and influence upon future naval wars. With twenty lithographic plates and many wood-cuts. 8vo, cloth............ 5 00

BEILSTEIN (F.) An Introduction to Qualitative Chemical Analysis, translated by I. J. Osbun. 12mo, cloth.......... 75

BENET (Gen. S. V., U. S. A.) Electro-Ballistic Machines, and the Schultz Chronoscope. Illustrated. Second edition, 4to, cloth 3 00

BLAKE (W. P.) Report upon the Precious Metals: Being Statistical Notices of the principal Gold and Silver producing regions of the World, represented at the Paris Universal Exposition. 8vo, cloth..................................... 2 00

—— Ceramic Art. A Report on Pottery, Porcelain, Tiles, Terra Cotta, and Brick. 8vo, cloth........................ 2 00

BOW (R. H.) A Treatise on Bracing, with its application to Bridges and other Structures of Wood or Iron. 156 illustrations. 8vo, cloth.... .... 1 50

BOWSER (Prof. E. A.) An Elementary Treatise on Analytic Geometry, embracing Plane Geometry, and an Introduction to Geometry of three Dimensions. 12mo, cloth....... 1 75

—— An Elementary Treatise on the Differential and Integral Calculus. With numerous examples. 12mo, cloth......... 2 25

BURGH (N. P.) Modern Marine Engineering, applied to Paddle and Screw Propulsion. Consisting of 36 colored plates, 259 practical wood-cut illustrations, and 403 pages of descriptive matter, the whole being an exposition of the present practice of James Watt & Co., J. & G. Rennie, R. Napier & Sons, and other celebrated firms. Thick 4to vol., cloth 10 00
Half morocco........ ....15 00

BURT (W. A.) Key to the Solar Compass, and Surveyor's Companion; comprising all the rules necessary for use in the field; also description of the Linear Surveys and Public Land System of the United States, Notes on the Barometer, suggestions for an outfit for a survey of four months, etc. Second edition. Pocket-book form, tuck............. 2 50

BUTLER (Capt. J. S., U. S. A.) Projectiles and Rifled Cannon. A Critical Discussion of the Principal Systems of Rifling and Projectiles, with Practical Suggestions for their Improvement, as embraced in a Report to the Chief of Ordnance, U. S. A. 36 plates. 4to, cloth....................... 6 00

CAIN (Prof. WM.) A Practical Treatise on Voussoir and Solid and Braced Arches. 16mo, cloth extra $1 75

CALDWELL (Prof. GEO. C.) and BRENEMAN (Prof. A. A.) Manual of Introductory Chemical Practice, for the use of Students in Colleges and Normal and High Schools. Third edition, revised and corrected. 8vo, cloth, illustrated. New and enlarged edition. 1 50

CAMPIN (FRANCIS). On the Construction of Iron Roofs. 8vo, with plates, cloth................. .... 2 00

CHAUVENET (Prof. W.) New method of correcting Lunar Distances, and improved method of finding the error and rate of a chronometer, by equal altitudes. 8vo, cloth...... 2 00

CHURCH (JOHN A.) Notes of a Metallurgical Journey in Europe. 8vo, cloth............ .... 2 00

CLARK (D. KINNEAR, C.E.) Fuel: Its Combustion and Economy, consisting of Abridgments of Treatise on the Combustion of Coal and the Prevention of Smoke, by C. W. Williams; and the Economy of Fuel, by T. S. Prideaux. With extensive additions on recent practice in the Combustion and Economy of Fuel: Coal, Coke, Wood, Peat, Petroleum, etc. 12mo, cloth... ....:............ 1 50

—— A Manual of Rules, Tables, and Data for Mechanical Engineers. Based on the most recent investigations. Illustrated with numerous diagrams. 1,012 pages. 8vo, cloth... 7 50
Half morocco............................ ....10 00

CLARK (Lt. LEWIS, U. S. N.) Theoretical Navigation and Nautical Astronomy. Illustrated with 41 wood-cuts. 8vo, cloth 1 50

CLARKE (T. C.) Description of the Iron Railway Bridge over the Mississippi River at Quincy, Illinois. Illustrated with 21 lithographed plans. 4to, cloth 7 50

CLEVENGER (S. R.) A Treatise on the Method of Government Surveying, as prescribed by the U. S. Congress and Commissioner of the General Land Office, with complete Mathematical, Astronomical, and Practical Instructions for the use of the United States Surveyors in the field. 16mo, morocco 2 50

COFFIN (Prof J. H. C.) Navigation and Nautical Astronomy. Prepared for the use of the U. S. Naval Academy. Sixth edition. 52 wood-cut illustrations. 12mo, cloth...... 3 50

COLBURN (ZERAH). The Gas-Works of London. 12mo, boards....... 60

COLLINS (JAS. E.) The Private Book of Useful Alloys and Memoranda for Goldsmiths, Jewellers, etc. 18mo, cloth... 50

CORNWALL (Prof. H. B.) Manual of Blow-Pipe Analysis, Qualitative and Quantitative, with a Complete System of Descriptive Mineralogy. 8vo, cloth, with many illustrations. N. Y., 1882 .... $2 50

CRAIG (B. F.) Weights and Measures. An account of the Decimal System, with Tables of Conversion for Commercial and Scientific Uses. Square 32mo, limp cloth.......... 50

CRAIG (Prof. THOS.) Elements of the Mathematical Theory of Fluid Motion. 16mo, cloth........................ 1 25

DAVIS (C. B.) and RAE (F. B.) Hand-Book of Electrical Diagrams and Connections. Illustrated with 32 full-page illustrations. Second edition. Oblong 8vo, cloth extra 2 00

DIEDRICH (JOHN). The Theory of Strains: a Compendium for the Calculation and Construction of Bridges, Roofs, and Cranes. Illustrated by numerous plates and diagrams. 8vo, cloth.. 5 00

DIXON (D. B.) The Machinist's and Steam-Engineer's Practical Calculator. A Compilation of useful Rules, and Problems Arithmetically Solved, together with General Information applicable to Shop-Tools, Mill-Gearing, Pulleys and Shafts, Steam-Boilers and Engines. Embracing Valuable Tables, and Instruction in Screw-cutting, Valve and Link Motion, etc. 16mo, full morocco, pocket form ...(In press)

DODD (GEO.) Dictionary of Manufactures, Mining, Machinery, and the Industrial Arts. 12mo, cloth............ 1 50

DOUGLASS (Prof S. H.) and PRESCOTT (Prof. A. B.) Qualitative Chemical Analysis. A Guide in the Practical Study of Chemistry, and in the Work of Analysis. Third edition. 8vo, cloth.. 3 50

DUBOIS (A. J.) The New Method of Graphical Statics. With 60 illustrations. 8vo, cloth 1 50

EASSIE (P. B.) Wood and its Uses. A Hand-Book for the use of Contractors, Builders, Architects, Engineers, and Timber Merchants. Upwards of 250 illustrations. 8vo, cloth 1 50

EDDY (Prof. H. T.) Researches in Graphical Statics, embracing New Constructions in Graphical Statics, a New General Method in Graphical Statics, and the Theory of Internal Stress in Graphical Statics. 8vo, cloth..................... 1 50

ELIOT (Prof. C. W.) and STORER (Prof. F. H.) A Compendious Manual of Qualitative Chemical Analysis. Revised with the co-operation of the authors. By Prof. William R. Nichols. Illustrated. 12mo, cloth................ .... 1 50

ELLIOT (Maj. GEO. H., U. S. E.) European Light-House Systems. Being a Report of a Tour of Inspection made in 1873. 51 engravings and 21 wood-cuts. 8vo, cloth 5 00

D. VAN NOSTRAND'S PUBLICATIONS. 5

ENGINEERING FACTS AND FIGURES. An Annual Register of Progress in Mechanical Engineering and Construction for the years 1863-64-65-66-67-68. Fully illustrated. 6 vols. 18mo, cloth (each volume sold separately), per vol..$2 50

FANNING (J. T.) A Practical Treatise of Water-Supply Engineering. Relating to the Hydrology, Hydrodynamics, and Practical Construction of Water-Works in North America. Third edition. With numerous tables and 180 illustrations. 650 pages. 8vo, cloth.................................... 5 00

FISKE (BRADLEY A., U. S. N.) Electricity in Theory and Practice. 8vo, cloth.. 2 50

FOSTER (Gen. J. G., U. S. A.) Submarine Blasting in Boston Harbor, Massachusetts. Removal of Tower and Corwin Rocks. Illustrated with seven plates. 4to, cloth.......... 3 50

FOYE (Prof. J. C.) Chemical Problems. With brief Statements of the Principles involved. Second edition, revised and enlarged. 16mo, boards..................................... 50

FRANCIS (JAS. B., C E.) Lowell Hydraulic Experiments: Being a selection from Experiments on Hydraulic Motors, on the Flow of Water over Weirs, In Open Canals of Uniform Rectangular Section, and through submerged Orifices and diverging Tubes. Made at Lowell, Massachusetts. Fourth edition, revised and enlarged, with many new experiments, and illustrated with twenty-three copperplate engravings. 4to, cloth 15 00

FREE-HAND DRAWING. A Guide to Ornamental Figure and Landscape Drawing. By an Art Student. 18mo, boards... 50

GILLMORE (Gen. Q. A.) Treatise on Limes, Hydraulic Cements, and Mortars. Papers on Practical Engineering, U. S. Engineer Department, No. 9, containing Reports of numerous Experiments conducted in New York City during the years 1858 to 1861, inclusive. With numerous illustrations. 8vo, cloth.. 4 00

—— Practical Treatise on the Construction of Roads, Streets, and Pavements. With 70 illustrations. 12mo, cloth....... 2 00

—— Report on Strength of the Building Stones in the United States, etc. 8vo, illustrated, cloth 1 50

—— Coignet Beton and other Artificial Stone. 9 plates, views, etc. 8vo, cloth... 2 50

GOODEVE (T. M.) A Text-Book on the Steam-Engine. 143 illustrations. 12mo, cloth....................................... 2 00

GORDON (J. E. H.) Four Lectures on Static Induction. 12mo, cloth... 80

D. VAN NOSTRAND'S PUBLICATIONS.

GRUNER (M. L.) The Manufacture of Steel. Translated from the French, by Lenox Smith, with an appendix on the Bessemer process in the United States, by the translator. Illustrated. 8vo, cloth.................................... $3 50

HALF-HOURS WITH MODERN SCIENTISTS. Lectures and Essays. By Professors Huxley, Barker, Stirling, Cope, Tyndall, Wallace, Roscoe, Huggins, Lockyer, Young, Mayer, and Reed. Being the University Series bound up. With a general introduction by Noah Porter, President of Yale College. 2 vols. 12mo, cloth, illustrated............. 2 50

HAMILTON (W. G.) Useful Information for Railway Men. Sixth edition, revised and enlarged 562 pages, pocket form. Morocco, gilt... 2 00

HARRISON (W. B.) The Mechanic's Tool Book, with Practical Rules and Suggestions for Use of Machinists, Iron-Workers, and others. Illustrated with 44 engravings. 12mo, cloth... 1 50

HASKINS (C. H.) The Galvanometer and its Uses. A Manual for Electricians and Students. Second edition. 12mo, morocco... 1 50

HENRICI (OLAUS). Skeleton Structures, especially in their application to the Building of Steel and Iron Bridges. With folding plates and diagrams. 8vo, cloth.................... 1 50

HEWSON (WM.) Principles and Practice of Embanking Lands from River Floods, as applied to the Levees of the Mississippi. 8vo, cloth....................................... 2 00

HOLLEY (ALEX. L.) A Treatise on Ordnance and Armor, embracing descriptions, discussions, and professional opinions concerning the materials, fabrication, requirements, capabilities, and endurance of European and American Guns, for Naval, Sea-Coast, and Iron-Clad Warfare, and their Rifling, Projectiles, and Breech-Loading; also, results of experiments against armor, from official records, with an appendix referring to Gun-Cotton, Hooped Guns, etc., etc. 948 pages, 493 engravings, and 147 Tables of Results, etc. 8vo, half roan... 10 00

—— Railway Practice American and European Railway Practice in the economical Generation of Steam, including the Materials and Construction of Coal-burning Boilers, Combustion, the Variable Blast, Vaporization, Circulation, Superheating, Supplying and Heating Feed-water, etc., and the Adaptation of Wood and Coke-burning Engines to Coal-burning; and in Permanent Way, including Road-bed, Sleepers, Rails, Joint-fastenings, Street Railways, etc., etc. With 77 lithographed plates. Folio, cloth.................12 00

HOWARD (C. R.) Earthwork Mensuration on the Basis of the Prismoidal Formulæ. Containing simple and labor-saving method of obtaining Prismoidal Contents directly

D. VAN NOSTRAND'S PUBLICATIONS.   7

from End Areas. Illustrated by Examples, and accompanied by Plain Rules for Practical Uses. Illustrated. 8vo, cloth $1 50

INDUCTION-COILS. How Made and How Used. 63 illustrations. 16mo, boards 50

ISHERWOOD (B. F.) Engineering Precedents for Steam Machinery. Arranged in the most practical and useful manner for Engineers. With illustrations. Two volumes in one. 8vo, cloth........ .... 2 50

JANNETTAZ (EDWARD). A Guide to the Determination of Rocks: being an Introduction to Lithology. Translated from the French by G. W. Plympton, Professor of Physical Science at Brooklyn Polytechnic Institute. 12mo, cloth.... 1 50

JEFFERS (Capt. W. N., U. S. N.) Nautical Surveying. Illustrated with 9 copperplates and 31 wood-cut illustrations. 8vo, cloth.... .... 5 00

JONES (H. CHAPMAN). Text-Book of Experimental Organic Chemistry for Students. 18mo, cloth.. .... 1 00

JOYNSON (F. H.) The Metals used in Construction: Iron, Steel, Bessemer Metal, etc., etc. Illustrated. 12mo, cloth. 75

—— Designing and Construction of Machine Gearing. Illustrated 8vo, cloth......... .... 2 00

KANSAS CITY BRIDGE (THE). With an account of the Regimen of the Missouri River, and a description of the methods used for Founding in that River. By O. Chanute, Chief-Engineer, and George Morrison, Assistant-Engineer. Illustrated with five lithographic views and twelve plates of plans   4to, cloth. 6 00

KING (W. H.) Lessons and Practical Notes on Steam, the Steam-Engine, Propellers, etc., etc, for young Marine Engineers, Students, and others. Revised by Chief-Engineer J. W. King, U. S. Navy. Nineteenth edition, enlarged. 8vo, cloth..... .... 2 00

KIRKWOOD (JAS. P.) Report on the Filtration of River Waters for the supply of Cities, as practised in Europe, made to the Board of Water Commissioners of the City of St. Louis. Illustrated by 30 double-plate engravings. 4to, cloth 15 00

LARRABEE (C. S.) Cipher and Secret Letter and Telegraphic Code, with Hogg's Improvements. The most perfect secret code ever invented or discovered. Impossible to read without the key. 18mo, cloth........ .... 1 00

LOCK (C. G.), WIGNER (G W.), and HARLAND (R. H.) Sugar Growing and Refining. Treatise on the Culture of Sugar-Yielding Plants, and the Manufacture and Refining of Cane, Beet, and other sugars. 8vo, cloth, illustrated 12 00

8 D. VAN NOSTRAND'S PUBLICATIONS.

LOCKWOOD (THOS. D.) Electricity, Magnetism, and Electro-Telegraphy. A Practical Guide for Students, Operators, and Inspectors. 8vo, cloth................................. $2 50

LORING (A. E.) A Hand-Book on the Electro-Magnetic Telegraph. Paper boards.. 50
Cloth.. 75
Morocco... 1 00

MAcCORD (Prof. C. W.) A Practical Treatise on the Slide-Valve by Eccentrics, examining by methods the action of the Eccentric upon the Slide-Valve, and explaining the practical processes of laying out the movements, adapting the valve for its various duties in the steam-engine. Second edition Illustrated. 4to, cloth 2 50

McCULLOCH (Prof. R S.) Elementary Treatise on the Mechanical Theory of Heat, and its application to Air and Steam Engines. 8vo, cloth..................................... 3 50

MERRILL (Col. WM. E, U. S. A.) Iron Truss Bridges for Railroads. The method of calculating strains in Trusses, with a careful comparison of the most prominent Trusses, in reference to economy in combination, etc., etc. Illustrated. 4to, cloth.. 5 00

MICHAELIS (Capt. O. E., U. S. A.) The Le Boulenge Chronograph, with three lithograph folding plates of illustrations. 4to, cloth... 3 00

MICHIE (Prof. P. S.) Elements of Wave Motion relating to Sound and Light. Text-Book for the U.S. Military Academy. 8vo, cloth, illustrated... 5 00

MINIFIE (WM.) Mechanical Drawing. A Text-Book of Geometrical Drawing for the use of Mechanics and Schools, in which the Definitions and Rules of Geometry are familiarly explained; the Practical Problems are arranged, from the most simple to the more complex, and in their description technicalities are avoided as much as possible. With illustrations for Drawing Plans, Sections, and Elevations of Railways and Machinery; an Introduction to Isometrical Drawing, and an Essay on Linear Perspective and Shadows. Illustrated with over 200 diagrams engraved on steel. Ninth edition. With an Appendix on the Theory and Application of Colors. 8vo, cloth 4 00

"It is the best work on Drawing that we have ever seen, and is especially a text-book of Geometrical Drawing for the use of Mechanics and Schools. No young Mechanic, such as a Machinist, Engineer, Cabinet-maker, Millwright, or Carpenter, should be without it."—*Scientific American.*

—— Geometrical Drawing. Abridged from the octavo edition, for the use of schools. Illustrated with forty-eight steel plates. Fifth edition. 12mo, cloth 2 00

MODERN METEOROLOGY. A Series of Six Lectures, delivered under the auspices of the Meteorological Society in 1878. Illustrated. 12mo, cloth.... ....$1 50

MORRIS (E.) Easy Rules for the Measurement of Earthworks, by Means of the Prismoidal Formula. 78 illustrations. 8vo, cloth.. 1 50

MOTT (H. A., Jr.) A Practical Treatise on Chemistry (Qualitative and Quantitative Analysis), Stoichiometry, Blow-pipe Analysis, Mineralogy, Assaying, Pharmaceutical Preparations, Human Secretions, Specific Gravities, Weights and Measures, etc., etc., etc. New edition, 1883. 650 pages. 8vo, cloth.. 4 00

NAQUET (A.) Legal Chemistry. A Guide to the Detection of Poisons, Falsification of Writings, Adulteration of Alimentary and Pharmaceutical Substances, Analysis of Ashes, and examination of Hair, Coins, Arms, and Stains, as applied to Chemical Jurisprudence, for the use of Chemists, Physicians, Lawyers, Pharmacists, and Experts. Translated, with additions, including a list of books and Memoirs on Toxicology, etc., from the French. By J. P. Battershall, Ph.D., with a preface by C. F. Chandler, Ph.D., M.D., LL.D. 12mo, cloth........ .... 2 00

NOBLE (W. H.) Useful Tables. Pocket form, cloth......... 50

NUGENT (E.) Treatise on Optics; or, Light and Sight, theoretically and practically treated, with the application to Fine Art and Industrial Pursuits. With 103 illustrations. 12mo, cloth..... .... 1 50

PEIRCE (B.) System of Analytic Mechanics. 4to, cloth.....10 00

PLANE TABLE (THE). Its Uses in Topographical Surveying. From the Papers of the U. S. Coast Survey. Illustrated. ..c., cloth............. .... 2 00
"This work gives a description of the Plane Table employed at the U. S. Coast Survey office, and the manner of using it."

PLATTNER. Manual of Qualitative and Quantitative Analysis with the Blow-Pipe. From the last German edition, revised and enlarged. By Prof. Th. Richter, of the Royal Saxon Mining Academy. Translated by Prof. H. B. Cornwall, assisted by John H. Caswell. Illustrated with 87 woodcuts and one lithographic plate. Fourth edition, revised, 560 pages. 8vo, cloth.. 5 00

PLYMPTON (Prof. GEO. W.) The Blow-Pipe. A Guide to its use in the Determination of Salts and Minerals. Compiled from various sources. 12mo, cloth........................ 1 50

—— The Aneroid Barometer: Its Construction and Use. Compiled from several sources. 16mo, boards, illustrated, 50
Morocco 1 00

PLYMPTON (Prof. GEO. W.) The Star-Finder, or Planisphere, with Movable Horizon Printed in colors on fine card-board, and in accordance with Proctor's Star Atlas... **$1 00**

POCKET LOGARITHMS, to Four Places of Decimals, including Logarithms of Numbers, and Logarithmic Sines and Tangents to Single Minutes. To which is added a Table of Natural Sines, Tangents, and Co-Tangents. 16mo, boards, **50**
Morocco **1 00**

POOK (S. M.) Method of Comparing the Lines and Draughting Vessels propelled by Sail or Steam. Including a chapter on Laying-off on the Mould-Loft Floor. 1 vol. 8vo, with illustrations, cloth **5 00**

POPE (F. L.) Modern Practice of the Electric Telegraph. A Hand-Book for Electricians and Operators. Eleventh edition, revised and enlarged, and fully illustrated. 8vo, cloth. **2 00**

PRESCOTT (Prof. A. B.) Outlines of Proximate Organic Analysis, for the Identification, Separation, and Quantitative Determination of the more commonly occurring Organic Compounds. 12mo, cloth **1 75**

—— Chemical Examination of Alcoholic Liquors. A Manual of the Constituents of the Distilled Spirits and Fermented Liquors of Commerce, and their Qualitative and Quantitative Determinations. 12mo, cloth **1 50**

—— First Book in Qualitative Chemistry. Second edition. 12mo, cloth **1 50**

PYNCHON (Prof. T. R.) Introduction to Chemical Physics, designed for the use of Academies, Colleges, and High-Schools. Illustrated with numerous engravings, and containing copious experiments with directions for preparing them. New edition, revised and enlarged, and illustrated by 269 illustrations on wood. Crown 8vo, cloth **3 00**

RAMMELSBERG (C. F.) Guide to a Course of Quantitative Chemical Analysis, especially of Minerals and Furnace Products. Illustrated by Examples. Translated by J. Towler, M.D. 8vo, cloth **2 25**

RANDALL (P. M.) Quartz Operator's Hand-Book. New edition, revised and enlarged, fully illustrated. 12mo, cloth... **2 00**

RANKINE (W. J. M.) Applied Mechanics, comprising Principles of Statics, Cinematics, and Dynamics, and Theory of Structures, Mechanism, and Machines. Crown 8vo, cloth. Tenth edition. London **5 00**

—— A Manual of the Steam-Engine and other Prime Movers, with numerous tables and illustrations. Crown 8vo, cloth. Tenth edition. London, 1882. .... **5 00**

—— A Selection from the Miscellaneous Scientific Papers of, with Memoir by P. G. Tait, and edited by W. J. Millar, C.E. 8vo, cloth. London, 1880 .... **10 00**

RANKINE (W. J. M.) A Manual of Machinery and Mill-work. Fourth edition. Crown 8vo. London, 1881 .... $5 00

―― Civil Engineering, comprising Engineering Surveys, Earthwork, Foundations, Masonry, Carpentry, Metal-works, Roads, Railways, Canals, Rivers, Water-works, Harbors, etc., with numerous tables and illustrations. Fourteenth edition, revised by E. F. Bamber, C.E. 8vo. London, 1883.... .... 6 50

―― Useful Rules and Tables for Architects, Builders, Carpenters, Coachbuilders, Engineers, Founders, Mechanics, Shipbuilders, Surveyors, Typefounders, Wheelwrights, etc. Sixth edition. Crown 8vo, cloth. London, 1883...... 4 00

―― and BAMBER (E. F.) A Mechanical Text-Book; or, Introduction to the Study of Mechanics and Engineering. 8vo, cl th. London, 1875 .... 3 50

RICE (Prof J. M.) and JOHNSON (Prof. W. W.) On a New Method of Obtaining the Differentials of Functions, with especial reference to the Newtonian Conception of Rates or Velocities. 12mo, paper 50

ROGERS (Prof. H. D.) The Geology of Pennsylvania. A Government Survey, with a General View of the Geology of the United States, Essays on the Coal Formation and its Fossils, and a description of the Coal Fields of North America and Great Britain. Illustrated with Plates and Engravings in the text. 3 vols. 4to, cloth, with Portfolio of Maps. 30 00

ROEBLING (J A) Long and Short Span Railway Bridges. Illustrated with large copperplate engravings of plans and views. Imperial folio, cloth 25 00

ROSE (JOSHUA, M.E.) The Pattern-Maker's Assistant, embracing Lathe Work, Branch Work Core Work, Sweep Work, and Practical Gear Constructions, the Preparation and Use of Tools, together with a large collection of useful and valuable Tables. Third edition. Illustrated with 250 engravings. 8vo, cloth 2 50

SABINE (ROBERT). History and Progress of the Electric Telegraph, with descriptions of some of the apparatus. Second edition, with additions, 12mo, cloth 1 25

SAELTZER (ALEX) Treatise on Acoustics in connection with Ventilation 12mo, cloth 1 00

SCHUMANN (F) A Manual of Heating and Ventilation in its Practical Application for the use of Engineers and Architects, embracing a series of Tables and Formulæ for dimensions of heating, flow and return pipes for steam and hot-water boilers, flues, etc, etc. 12mo. Illustrated. Full roan . .. 1 50

―― Formulas and Tables for Architects and Engineers in calculating the strains and capacity of structures in Iron and Wood 12mo, morocco, tucks 2 50

12 D. VAN NOSTRAND'S PUBLICATIONS.

SAWYER (W. E.) Electric-Lighting by Incandescence, and its Application to Interior Illumination. A Practical Treatise. With 96 illustrations. Third edition. 8vo, cloth.$2 50

SCRIBNER (J. M.) Engineers' and Mechanics' Companion, comprising United States Weights and Measures, Mensuration of Superfices and Solids, Tables of Squares and Cubes, Square and Cube Roots, Circumference and Areas of Circles, the Mechanical Powers, Centres of Gravity, Gravitation of Bodies, Pendulums, Specific Gravity of Bodies, Strength, Weight, and Crush of Materials, Water-Wheels, Hydrostatics, Hydraulics, Statics, Centres of Percussion and Gyration, Friction Heat, Tables of the Weight of Metals, Scantling, etc., Steam and the Steam-Engine. Nineteenth edition, revised, 16mo, full morocco............ 1 50

—— Engineers', Contractors', and Surveyors' Pocket Table-Book. Comprising Logarithms of Numbers, Logarithmic Sines and Tangents, Natural Sines and Natural Tangents, the Traverse Table, and a full and complete set of Excavation and Embankment Tables, together with numerous other valuable tables for Engineers, etc. Eleventh edition, revised, 16mo, full morocco 1 50

SHELLEN (Dr. H.) Dynamo-Electric Machines. Translated, with much new matter on American practice, and many illustrations which now appear for the first time in print. 8vo, cloth, New York............................(In press)

SHOCK (Chief-Eng. W. H.) Steam-Boilers: their Design, Construction, and Management. 450 pages text. Illustrated with 150 wood-cuts and 36 full-page plates (several double). Quarto. Illustrated. Half morocco..................... 15 00

SHUNK (W. F.) The Field Engineer. A handy book of practice in the Survey, Location, and Track-work of Railroads, containing a large collection of Rules and Tables, original and selected, applicable to both the Standard and Narrow Gauge, and prepared with special reference to the wants of the young Engineer. Third edition. 12mo, morocco, tucks... 2 50

SHIELDS (J. E.) Notes on Engineering Construction. Embracing Discussions of the Principles involved, and Descriptions of the Material employed in Tunnelling, Bridging, Canal and Road Building, etc., etc. 12mo, cloth 1 50

SHREVE (S. H) A Treatise on the Strength of Bridges and Roofs. Comprising the determination of Algebraic formulas for strains in Horizontal, Inclined or Rafter, Triangular, Bowstring, Lenticular, and other Trusses, from fixed and moving loads, with practical applications and examples, for the use of Students and Engineers. 87 wood-cut illustrations. Third edition. 8vo, cloth........................... 3 50

SIMMS (F. W.) A Treatise on the Principles and Practice of Levelling; showing its application to purposes of Railway Engineering and the Construction of Roads, etc. Revised and corrected, with the addition of Mr. Laws's Practical Examples for setting out Railway Curves. Illustrated. 8vo, cloth $2 50

STILLMAN (PAUL.) Steam-Engine Indicator, and the Improved Manometer Steam and Vacuum Gauges—their Utility and Application. New edition. 12mo, flexible cloth 1 00

STONEY (B. D.) The Theory of Strains in Girders and similar structures, with observations on the application of Theory to Practice, and Tables of Strength and other properties of Materials. New and revised edition, enlarged. Royal 8vo, 664 pages. Complete in one volume. 8vo, cloth........12 50

STUART (CHAS. B., U. S. N.) The Naval Dry Docks of the United States. Illustrated with 24 engravings on steel. Fourth edition, cloth 6 00

—— The Civil and Military Engineers of America. With 9 finely executed portraits of eminent engineers, and illustrated by engravings of some of the most important works constructed in America. 8vo, cloth....................... 5 00

STUART (B.) How to Become a Successful Engineer. Being Hints to Youths intending to adopt the Profession. Sixth edition. 12mo, boards 50

SWEET (S. H.) Special Report on Coal, showing its Distribution, Classification, and Cost delivered over different routes to various points in the State of New York and the principal cities on the Atlantic Coast. With maps. 8vo, cloth 3 00

TEXT-BOOK (A) ON SURVEYING, Projections, and Portable Instruments, for the Use of the Cadet Midshipmen at the U. S. Naval Academy. Nine lithographed plates and several wood-cuts. 8vo, cloth 2 00

TONER (J. M.) Dictionary of Elevations and Climatic Register of the United States. Containing, in addition to Elevations, the Latitude, Mean Annual Temperature, and the total Annual Rain-fall of many localities; with a brief introduction on the Orographic and Physical Peculiarities of North America. 8vo, cloth 3 75

TUCKER (Dr. J. H.) A Manual of Sugar Analysis, including the Applications in General of Analytical Methods to the Sugar Industry. With an Introduction on the Chemistry of Cane Sugar, Dextrose, Levulose, and Milk Sugar. 8vo, cloth, illustrated.. 3 50

TUNNER (P.) A Treatise on Roll-Turning for the Manufacture of Iron Translated and adapted by John B. Pearse, of the Pennsylvania Steel-Works, with numerous engravings, wood-cuts, and folio atlas of plates................10 00

VAN WAGENEN (T. F.) Manual of Hydraulic Mining, for the use of the Practical Miner. 12mo, cloth.................$1 00

WALKER (W. H.) Screw Propulsion. Notes on Screw Propulsion: Its Rise and History. 8vo, cloth. 75

WANKLYN (J. A.) A Practical Treatise on the Examination of Milk and its Derivatives, Cream, Butter, and Cheese. 12mo, cloth................. .... 1 00

WATT (ALEX.) Electro-Metallurgy, Practically Treated. Sixth edition, with considerable additions   12mo, cloth.... 1 00

WEISBACH (JULIUS). A Manual of Theoretical Mechanics. Translated from the fourth augmented and improved German edition, with an introduction to the Calculus, by Eckley B. Coxe, A.M., Mining Engineer. 1,100 pages, and 902 wood-cut illustrations. 8vo, cloth........................10 00

WEYRAUCH (J. J.) Strength and Calculations of Dimensions of Iron and Steel Construction, with reference to the Latest Experiments. 12mo, cloth, plates............. 1 00

WILLIAMSON (R. S) On the use of the Barometer on Surveys and Reconnoissances. Part I. Meteorology in its Connection with Hypsometry. Part II. Barometric Hypsometry. With Illustrative Tables and Engravings. 4to. cloth 15 00

—— Practical Tables in Meteorology and Hypsometry, in connection with the use of the Barometer. 4to, cloth 2 50

Complete 112-page Catalogue of works in every department of science sent postpaid to any address on receipt of ten cents in postage stamps.

THE UNIVERSITY SERIES.

No. 1. ON THE PHYSICAL BASIS OF LIFE. By Prof. T. H. Huxley, LL.D., F.R.S. With an introduction by a Professor in Yale College. 12mo, pp. 36. Paper cover.............. 25

No. 2. THE CORRELATION OF VITAL AND PHYSICAL FORCES. By Prof. George F. Barker, M.D., of Yale College. 36 pp. Paper covers................................. 25

No. 3. AS REGARDS PROTOPLASM, in relation to Prof. Huxley's "Physical Basis of Life." By J. Hutchinson Stirling, F.R.C.S. 72 pp.... 25

No. 4. ON THE HYPOTHESIS OF EVOLUTION, Physical and Metaphysical. By Prof. Edward D. Cope. 12mo, 72 pp. Paper covers. 25

No. 5. SCIENTIFIC ADDRESSES—1, On the Methods and Tendencies of Physical Investigation   2. On Haze and Dust. 3 On the Scientific Use of the Imagination. By Prof. John Tyndall, F.R.S. 12mo, 74 pp. Paper covers..... 25
Flex. cloth..... 50

No. 6. NATURAL SELECTION AS APPLIED TO MAN. By Alfred Russell Wallace. This pamphlet treats (1) of the Development of Human Races under the Law of Selection; (2) the Limits of Natural Selection as applied to Man. 54 pp. 25

No. 7. SPECTRUM ANALYSIS. Three lectures by Profs. Roscoe, Huggins, and Lockyer. Finely illustrated. 88 pp. Paper covers 25

No. 8. THE SUN. A sketch of the present state of scientific opinion as regards this body. By Prof. C. A. Young, Ph.D., of Dartmouth College. 58 pp. Paper covers............. .. 25

No. 9. THE EARTH A GREAT MAGNET. By A. M. Mayer, Ph.D., of Stevens Institute. 72 pp. Paper covers.......... 25
Flexible cloth 50

No. 10. MYSTERIES OF THE VOICE AND EAR. By Prof. O. N Rood, Columbia College, New York. Beautifully illustrated. 38 pp. Paper covers 25

Or together, 2 vols., cloth................. ....$2 50

VAN NOSTRAND'S SCIENCE SERIES.

(18mo, green boards. Amply illustrated where the subject demands.)

No. 1. CHIMNEYS FOR FURNACES, FIRE-PLACES, AND STEAM-BOILERS. By R. Armstrong, C.E. Second edition, enlarged........ 50
No. 2. STEAM-BOILER EXPLOSIONS. By Zerah Colburn.. 50
No. 3 PRACTICAL DESIGNING OF RETAINING WALLS. By Arthur Jacob, A.B. 50
No. 4. PROPORTIONS OF PINS USED IN BRIDGES. By Charles Bender C.E......... 50
No. 5. VENTILATION OF BUILDINGS. By W. F. Butler... 50
No. 6 ON THE DESIGNING AND CONSTRUCTION OF STORAGE RESERVOIRS. By Arthur Jacob, A.B 50
No 7. SURCHARGED AND DIFFERENT FORMS OF RETAINING WALLS. By James S. Tate, C.E.............. 50
No. 8. A TREATISE ON THE COMPOUND ENGINE. By John Turnbull, Jr. Second edition, revised by Prof. S. W. Robinson 50
No. 9. FUEL. By C. William Siemens, D C.L.; to which is appended the VALUE OF ARTIFICIAL FUEL AS COMPARED WITH COAL. By John Wormald, C.E 50
No 10. COMPOUND ENGINES. Translated from the French of A. Mallet 50
No. 11. THEORY OF ARCHES. By Prof. W. Allan. 50
No. 12. A THEORY OF VOUSSOIR ARCHES. By Prof. W. E. Cain 50
No. 13 GASES MET WITH IN COAL-MINES. By J. J. Atkinson 50
No. 14. FRICTION OF AIR IN MINES. By J. J. Atkinson ... 50
No. 15. SKEW ARCHES. By Prof. E. W. Hyde, C.E. Illustrated 50
No. 16. A GRAPHIC METHOD FOR SOLVING CERTAIN ALGEBRAIC EQUATIONS. By Prof. George L. Vose.. 50
No. 17. WATER AND WATER SUPPLY. By Prof. W. H. Corfield, of the University College, London............ 50
No. 18. SEWERAGE AND SEWAGE UTILIZATION. By Prof. W. H. Corfield, M.A., of the University College, London............. ;....... 50
No. 19. STRENGTH OF BEAMS UNDER TRANSVERSE LOADS. By Prof. W. Allan, author of "Theory of Arches" 50

No. 20. BRIDGE AND TUNNEL CENTRES. By John B. McMaster, C.E.. 50
No. 21. SAFETY VALVES. By Richard H. Buel, C.E........ 50
No. 22. HIGH MASONRY DAMS. By John B. McMaster, C.E.. 50
No. 23. THE FATIGUE OF METALS UNDER REPEATED STRAINS, with Various Tables of Results and Experiments. From the German of Prof. Ludwig Spangenburgh, with a Preface by S. H. Shreve, A.M................................ 50
No. 24. A PRACTICAL TREATISE ON THE TEETH OF WHEELS. By Prof. S. W. Robinson........................ 50
No. 25 ON THE THEORY AND CALCULATION OF CONTINUOUS BRIDGES. By Mansfield Merriman, Ph.D..... 50
No. 26. PRACTICAL TREATISE ON THE PROPERTIES OF CONTINUOUS BRIDGES. By Charles Bender, C.E. 50
No. 27. ON BOILER INCRUSTATION AND CORROSION. By F. J. Rowan..................................... 50
No 28. TRANSMISSION OF POWER BY WIRE ROPES. By Albert W. Stahl, U. S. N......................... 50
No. 29. STEAM INJECTORS. Translated from the French of M. Leon Pochet................................. 50
No. 30. TERRESTRIAL MAGNETISM AND THE MAGNETISM OF IRON VESSELS. By Prof. Fairman Rogers... 50
No. 31. THE SANITARY CONDITION OF DWELLING-HOUSES IN TOWN AND COUNTRY. By George E. Waring, Jr.. 50
No 32. CABLE-MAKING OF SUSPENSION BRIDGES. By W. Hildenbrand, C.E............................... 50
No. 33. MECHANICS OF VENTILATION. By George W. Rafter, C E....................................... 50
No. 34 FOUNDATIONS By Prof. Jules Gaudard, C.E. Translated from the French.............................. 50
No. 35. THE ANEROID BAROMETER: ITS CONSTRUCTION AND USE. Compiled by George W. Plympton Second edition.................................... 50
No. 36. MATTER AND MOTION. By J. Clerk Maxwell, M.A. 50
No. 37. GEOGRAPHICAL SURVEYING: Its Uses, Methods, and Results. By Frank De Yeaux Carpenter, C E. 50
No. 38. MAXIMUM STRESSES IN FRAMED BRIDGES. By Prof. Wm. Cain, A.M., C.E........................ 50
No 39. A HAND-BOOK OF THE ELECTRO-MAGNETIC TELEGRAPH. By A. E. Loring..................... 50
No. 40. TRANSMISSION OF POWER BY COMPRESSED AIR. By Robert Zahner, M.E....................... 50
No 41. STRENGTH OF MATERIALS. By Wm. Kent, C.E.. 50

No. 42. VOUSSOIR ARCHES APPLIED TO STONE BRIDGES, TUNNELS, CULVERTS, AND DOMES. By Prof. Wm. Cain 50

No. 43. WAVE AND VORTEX MOTION. By Dr. Thomas Craig, of Johns Hopkins University. 50

No. 44. TURBINE WHEELS. By. Prof. W. P. Trowbridge, Columbia College 50

No 45. THERMODYNAMICS. By Prof. H. T. Eddy, University of Cincinnati 50

No. 46. ICE-MAKING MACHINES. From the French of M. Le Doux 50

No. 47. LINKAGES; THE DIFFERENT FORMS AND USES OF ARTICULATED LINKS. By J. D. C. De Roos 50

No. 48. THEORY OF SOLID AND BRACED ARCHES. By Wm. Cain, C.E. 50

No. 49. ON THE MOTION OF A SOLID IN A FLUID. By Thomas Craig, Ph D 50

No. 50. DWELLING HOUSES: Their Sanitary Construction and Arrangements. By Prof. W. H. Corfield 50

No. 51. THE TELESCOPE: Its Construction, etc. By Thos. Nolan 50

No. 52 IMAGINARY QUANTITIES Translated from the French of M. Argano. By Prof. Hardy 50

No. 53. INDUCTION COILS: How Made and How Used 50

No. 54. KINEMATICS OF MACHINERY. By Prof. Kennedy. With an introduction by Prof. R. H. Thurston 50

No. 55. SEWER GASES: Their Nature and Origin. By A. De Varona 50

No. 56. THE ACTUAL LATERAL PRESSURE OF EARTH-WORK. By Benj. Baker, M. Inst. C.E 50

No. 57. INCANDESCENT ELECTRIC LIGHTS, WITH PARTICULAR REFERENCE TO THE EDISON LAMPS AT THE PARIS EXHIBITION. By Comte Th. Du Moncel, Wm. Henry Preece, J. W. Howell, and others. Second edition 50

No. 58. THE VENTILATION OF COAL-MINES. By W. Fairley, M.E., F.S.S 50

No. 59. RAILROAD ECONOMICS; or, Notes, with Comments. By S. W. Robinson, C.E 50

No. 60. STRENGTH OF WROUGHT-IRON BRIDGE MEMBERS. By S. W. Robinson, C E 50

No. 61. POTABLE WATER AND THE DIFFERENT METHODS OF DETECTING IMPURITIES By Chas. W. Folkard. 50

No. 62. THE THEORY OF THE GAS ENGINE. By Dugald Clerk..... 50
No. 63. HOUSE DRAINAGE AND SANITARY PLUMBING. By W. P. Gerhard.... 50
No. 64. ELECTRO-MAGNETS. By Th. Du Moncel..... 50
No. 65. POCKET LOGARITHMS TO FOUR PLACES DECIMALS..... 50
No. 66. DYNAMO-ELECTRIC MACHINERY. By S. P. Thompson. With notes by F. L. Pope..... 50
No. 67. HYDRAULIC TABLES, BASED ON "KUTTER'S FORMULA." By P. J. Flynn..... 50
No. 68. STEAM-HEATING. By Robert Briggs..... 50
No. 69. CHEMICAL PROBLEMS. By Prof. J. C. Foye. Second edition, revised and enlarged..... 50
No. 70. EXPLOSIVES AND EXPLOSIVE COMPOUNDS. By M. Bertholet..... 50
No. 71. DYNAMIC ELECTRICITY. By John Hopkinson, J. A. Schoolbred, and R. E. Day..... 50
No. 72. TOPOGRAPHICAL SURVEYING. By Geo. J. Specht, Prof. A. S. Hardy, John B. McMaster, and H. F. Walling..... 50
No. 73. SYMBOLIC ALGEBRA; or, The Algebra of Algebraic Numbers. By Prof. W. Cain..... 50
No. 74. TESTING-MACHINES, Their History, Construction, and Use. By Arthur V. Abbott..... 50
No. 75. RECENT PROGRESS IN DYNAMO-ELECTRIC MACHINES. Being a Supplement to Dynamo-Electric Machinery. By Prof. Sylvanus P. Thompson..... 50
No. 76. MODERN REPRODUCTIVE GRAPHIC PROCESSES. By Lt. Jas. S. Pettit, U.S.A..... 50
No. 77. STADIA SURVEYING. The Theory of Stadia Measurements. By Arthur Winslow..... 50
No. 78. THE STEAM-ENGINE INDICATOR, and Its Use. By W. B. Le Van..... 50
No. 79. THE FIGURE OF THE EARTH. By Frank C. Roberts, C.E..... 50
No. 80. HEALTHY FOUNDATIONS FOR HOUSES. By Glenn Brown..... 50
No. 81. WATER METERS: Comparative Tests of Accuracy, Delivery, etc. Distinctive features of the Worthington, Kennedy, Siemens, and Hesse Meters. By Ross E. Browne..... 50
No. 82. THE PRESERVATION OF TIMBER by the use of Antiseptics. By Samuel Bagster Boulton, C.E..... 50
No. 83. MECHANICAL INTEGRATORS. By Prof. Henry S. H. Shaw, C.E..... 50

www.ingramcontent.com/pod-product-compliance
Lightning Source LLC
Chambersburg PA
CBHW031604110426
42742CB00037B/1212